Manufacturing Engineering and Technology

Manufacturing Engineering and Technology

Joshua Holt

WILLFORD PRESS

www.willfordpress.com

Published by Willford Press,
118-35 Queens Blvd., Suite 400,
Forest Hills, NY 11375, USA

ISBN: 978-1-64728-334-6

Cataloging-in-Publication Data

Manufacturing engineering and technology / Joshua Holt.
   p. cm.
Includes bibliographical references and index.
ISBN 978-1-64728-334-6
1. Production engineering. 2. Manufacturing processes 3. Industrial engineering. 4. Manufactures--Technological innovations. 5. Technological innovations. 6. Industrial equipment. I. Holt, Joshua.
TS176 .M36 2022
670--dc23

For information on all Willford Press publications
visit our website at www.willfordpress.com

(CILLFORD PRESS**

TABLE OF CONTENTS

It is with great pleasure that I present this book. It has been carefully written after numerous discussions with my peers and other practitioners of the field. I would like to take this opportunity to thank my family and friends who have been extremely supporting at every step in my life.

Manufacturing engineering is a branch of professional engineering. It is closely associated with other disciplines of engineering such as mechanical, industrial and electrical engineering. It focuses on researching and developing tools, machines and equipment and planning the practices of manufacturing. It also focuses on the production of quality products with the optimum expenditure of capital. There are various sub-disciplines within this field such as mechanics, kinematics, drafting, machine tools and metal fabrication, computer integrated manufacturing and mechatronics. This book unfolds the innovative aspects of manufacturing engineering which will be crucial for the holistic understanding of the subject matter. While understanding the long-term perspectives of the topics, it makes an effort in highlighting their impact as a modern tool for the growth of the discipline. This textbook is an essential guide for both academicians and those who wish to pursue this discipline further.

The chapters below are organized to facilitate a comprehensive understanding of the subject:

Chapter – Manufacturing and its Types

Manufacturing is the process of producing products either for use or for the purpose of sale by the means of manpower, machines, tools or chemical or biological processes. It can be categorized into discrete manufacturing, project manufacturing, lean manufacturing, agile manufacturing, etc. The topics elaborated in this chapter will help in gaining a better perspective about manufacturing and its types.

Chapter – Manufacturing Engineering

The branch of engineering which focuses on developing tools, processes, machines, etc. and integrating them to maximize the revenue is called manufacturing engineering. It is mainly bifurcated into manufacturing production and packaging engineering. This chapter discusses about manufacturing engineering and its aspects in detail.

Chapter – Understanding Manufacturing Processes

Manufacturing processes comprise of numerous procedures such as mesoscale manufacturing, laser engraving, casting, lost-foam casting, permanent mold casting, molding, blow molding, thermoforming, lamination, dip molding, rotational molding, thermal spraying, etc. This chapter has been carefully written to provide an easy understanding of these manufacturing processes.

Chapter – Manufacturing Tools and Technologies

Manufacturing engineering requires numerous tools and technologies to perform various operations. Conveyor system, shaper, optical comparator, water jet cutter, lathe, jig, pattern, fixture, die, material-handling tools, etc. are some of the tools and technologies used. This chapter closely examines these tools and technologies used in manufacturing engineering to provide an extensive understanding of the subject.

Chapter – Digital Manufacturing

A consolidated approach to manufacturing which primarily makes use of computing systems is defined as digital manufacturing. Computer-integrated manufacturing, digital factory, cloud manufacturing, cyber manufacturing, digital materialization, distributed manufacturing, cloud-based designing, laser rapid manufacturing, etc. are some of its applications. All these diverse principles and applications of digital manufacturing have been carefully analyzed in this chapter.

Joshua Holt

Manufacturing and its Types

Manufacturing is the process of producing products either for use or for the purpose of sale by the means of manpower, machines, tools or chemical or biological processes. It can be categorized into discrete manufacturing, project manufacturing, lean manufacturing, agile manufacturing, etc. The topics elaborated in this chapter will help in gaining a better perspective about manufacturing and its types.

Manufacturing is the process of transforming materials or components into finished products that can be sold in the marketplace. Every physical product that you buy in a store or online is manufactured somewhere. The manufacturing industry is one of the largest sectors of the U.S. economy, employing more than 12 million workers. Today, technology is causing the country's economy to move towards offering services as opposed to producing goods. However, it is becoming clear to economists that a healthy manufacturing industry is one of the hallmark indicators of a healthy, thriving economy. And manufacturing intermingles with nearly every area of the economy.

Manufacturing Industries

Manufacturing industries are those that engage in the transformation of goods, materials or substances into new products. The transformational process can be physical, chemical or mechanical. Manufacturers often have plants, mills or factories that produce goods for public consumption. Machines and equipment are typically used in the process of manufacturing. Although, in some cases, goods can be manufactured by hand. An example of this would be baked goods, handcrafted jewelry, other handicrafts and art.

There are several massive manufacturing industries including food, beverage, tobacco, textiles, apparel, leather, paper, oil and coal, plastics and rubbers, metal, machinery, computers and electronics, transportation, furniture and others. More than 12 million Americans are employed across manufacturing industries. Further, many millions more are employed indirectly by the manufacturing industries. Manufacturing is vital to the U.S. economy, making up a large percentage of the country's gross domestic product (GDP). Manufacturing industries are responsible for the goods in our economy, or the physical products we buy and use every day.

What Manufacturing Industries do?

Manufacturers create physical goods. How these goods are created varies depending on the specific company and industry. However, most manufacturers use machinery and industrial equipment to produce goods for public consumption. The manufacturing process creates value, meaning companies can charge a premium for what they create. For example, rubber is not particularly valuable on its own. But when it is formed into a car tire, it holds substantially more value. So, in this case, the manufacturing process that allows the rubber to be transformed into a necessary car part adds value.

Prior to the Industrial Revolution, the majority of goods were made by hand. Since the Industrial Revolution, manufacturing has grown increasingly important, with many goods being massed produced. Mass production means that goods can be produced much more quickly and with more precision. This drives down prices and makes many consumer goods cheaper, their cost within reach of the general public. When the assembly line was introduced into manufacturing, production further skyrocketed. Then, in the early 20th century, Henry Ford introduced a conveyor belt that physically moved products through the factory, from one station to the next. Each station also had a worker responsible for fulfilling a specific stage in the production process. This simple conveyor belt tripled production, and changed manufacturing forever.

Today's advancement of computer technology allows manufacturers to do more with less time. Now, thousands of items can be manufactured within the space of minutes. Computer technology can be used to assemble, test and track production. Each year, technology continues to make manufacturing increasingly efficient, faster and more cost-effective. However, automation also eliminates many manufacturing jobs, leaving skilled employees without work.

Manufacturing Industries Examples

There are many manufacturing industries in the United States. The following are some of the most popular manufacturing sectors in the country:

- Food manufacturing: The food sector of manufacturing transforms agricultural or livestock products into products for consumption. Typically, these are sold to wholesalers or retailers who then sell those products to consumers. A few examples of food manufacturing products are baked goods, grains, fruit and vegetable preserves and animal meat.

- Beverage and tobacco product manufacturing: Interestingly, tobacco and beverages are in the same sector of manufacturing. Beverage products include those that are non-alcoholic, as well as those that are alcoholic through the fermentation or distillation process. Ice is also considered a manufactured beverage. Tobacco products are loose tobacco products, as well as those that are in cigarette or cigar form.

- Textile manufacturing: Textile manufacturers turn fibers into usable fabrics that will eventually be transformed into consumer goods such as apparel, sheets, towels or curtains. A few examples of textile manufacturing are fiber, yarn, thread and fabric mills.

- Apparel manufacturing: Apparel manufacturers fall into two main types. The first is cut and sew, meaning a garment is created by purchasing fabric, cutting it up and then sewing it. The second type of apparel manufacturing involves knitting the fabric and then cutting and sewing it. The apparel sector is extremely popular and encompasses many different kinds of workers, including tailors and even knitters.

- Leather and allied product manufacturing: This sector is concerned with the manufacturing of leather as well as leather substitutes such as rubbers or plastics. The reason leather substitutes fall under this sector of manufacturing is that they are often made in the same factories with the same machinery as leather products. It doesn't make sense for manufacturers to separate them, so they are both included.

- Wood product manufacturing: Wood manufacturing covers products like lumber, plywood, veneers, flooring and more. Further, manufactured homes and prefabricated wood buildings are considered wood product manufacturing. Wood must be cut, shaped and finished. Some manufacturers use logs to make their wood products while others purchase pre-cut lumber and further process the wood from there.

- Paper manufacturing: Paper manufacturers make pulp, paper or convert paper products. These three processes are grouped because many manufacturers do all three. It would be cumbersome to separate these activities from one another, so it makes sense to group them.

- Petroleum and coal manufacturing: This industry is concerned with transforming crude petroleum and coal into usable consumer products. Petroleum requires refining before consumers can use it. The refining process separates different components of the petroleum for different products.

- Chemical manufacturing: Chemicals manufacturing encompasses several different industries. This manufacturing process is the transformation of organic or inorganic materials into a unique product. A few examples of this are pesticides, fertilizers, pharmaceuticals, soaps, cleaning compounds and more.

- Plastics and rubbers manufacturing: This manufacturing sector makes rubbers and plastics. The two are lumped together because they are used as substitutes for one another. However, each is its own subsector, meaning plants can usually only manufacture one of the two; not both.

- Metal manufacturing: The metal manufacturing sector produces metals like iron, steel, aluminum and more. It also includes foundries.

- Fabricated metals: Under this sector, metals are transformed into other end products. Some examples of products are cutlery, hand tools, hardware, springs, screws, nuts and bolts.

- Machinery manufacturing: This sector of manufacturing creates machines that apply mechanical force. The machines are created through processes like forging, stamping, bending, forming, welding and the assembling of parts. Machinery manufacturing is complex and covers many processes. Machines are complicated and require many parts, not to mention specific mechanics. For example, a piece of industrial machinery could have a computer, as well as many other components. Machine manufacturing encompasses agricultural, construction, mining, heating, cooling, ventilation, air conditioning, refrigeration, engines and more.

- Computer and electronics manufacturing: This sector of manufacturing is rapidly growing and continues to grow. The insatiable demand for electronics makes this a highly competitive industry. Because of the use of integrated circuits and miniaturized technology, this is a specialized manufacturing sector. This grouping includes computers, communications equipment and audio and visual equipment, to name a few.

- Transportation equipment manufacturing: Nearly everything to do with the transporting of goods and people is produced in this manufacturing sector. This is a massive sector of the manufacturing industry, encompassing motor vehicles, planes, trains and ships. Transportation equipment, in general, qualifies as machinery. These manufacturing processes are extremely complex and require many different components being made in the same factories.

- Furniture manufacturing: This sector of the manufacturing industry includes furniture and all other related products such as mattresses, blinds, cabinets and lighting. Goods that are manufactured in this sector must be functional and have a well-thought-out design. There are countless processes that can go into manufacturing furniture. One such example is the cutting, shaping, finishing and attaching of wood to make a table.

Why Manufacturing Industries Matter

Manufacturing industries matter for several reasons. Historically, the United States has been one of the world's largest – if not the largest manufacturer of goods. The manufacturing and exporting of goods help keep money flowing into the U.S. economy. Economies thrive when they have strong manufacturing industries. Further, when manufacturing is thriving, innovation soars. Manufacturers produce roughly 75 percent of all privately funded research and development in the country. Manufacturing is a huge propeller of innovation and forward thinking.

Another reason manufacturing industries matter is because factory jobs tend to be middle-class jobs that pay above-average wages. Manufacturing is one of the few industries where a worker without an advanced degree can earn a living wage. Because it is one of the country's largest employment sectors, a lot of families rely on manufacturing industries to put food on the table. The industrial sector also supports many secondary industries. Manufacturing supports roughly 1-in-6 service jobs. Even manufacturing companies need lawyers, accountants, doctors, financial advisors and other service professionals.

Manufacturing industries also spur investments and encourage the building of infrastructure. There are few areas of the economy that manufacturing industries don't touch. Many other industries contribute directly and indirectly to manufacturing. A few examples are construction, engineering, printing and transportation, which are all needed to help keep manufacturing afloat. A new factory can't be built without an engineer, an architect and a construction crew. Clothing manufacturers can't get their products to stores without shipping their products. New products can't be developed without research and development teams, engineers and product designers. Countless companies would cease to exist without manufacturing, as they would have no products to sell. Ultimately, manufacturing industries are deeply entangled in the world economy.

It is unclear whether manufacturing will continue to decline., or whether it will begin to thrive again. There does not seem to be a consensus among economists. Some believe that we are moving into a post-goods economy where services will reign supreme. Others believe that manufacturing will continue to grow, though it will evolve with technology. Manufacturing jobs may become highly skilled technical jobs that require advanced training. Companies might hire engineers rather than blue collar workers. It's difficult to predict what will happen. However, what remains clear is that for now, manufacturing has an important role to play in both the economy and the labor force.

DISCRETE MANUFACTURING

Discrete manufacturing is where finished distinct products are created and assembled.

Almost every item sold in stores is an example of discrete manufacturing. What is meant by discrete manufacturing is that the object being created is a distinct unit. You can divide non-distinct products, like oil, into any size you want. You cannot divide a teapot into two halves because it is a distinct unit.

Examples of discrete manufacturing could include:

- Vehicles,

- Aircraft,

- Smartphones,

- Computers,

- Cookware,

- Clothing,

- Cabling.

It could also include component parts such as:

- Nuts,

- Bolts,

- Brackets.

These component parts can be individually countable as units, or identifiable as numbers. Usually once produced, items cannot be distilled back into original components.

Discrete manufacturing can be characterized by unit production; where units can be produced with high complexity and low volume, like aircrafts or computers, or low complexity and high volume, like nuts or bolts.

What is the Difference between Discrete and Process Manufacturing?

Process manufacturing relies on creating formulas or recipes to produce a product, whereas discrete manufacturing assembles parts in a prescribed process to produce a distinct item. While discrete manufacturing creates products that are differentiated by individual units, process manufacturing does not. While discrete manufacturing produces individual units, like smartphones, process manufacturing produces indistinct units, like salt, oil, and water.

The following products are good examples of process manufacturing:

- Food,

- Beverages,

- Pharmaceuticals,

- Chemicals,

- Plastics.

What is the Difference between Manufacturing and Assembly?

Manufacturing involves taking raw materials and using them to craft a part or component. A company may manufacture springs from hardened steel. Assembly is the process of taking parts, often made by manufacturing, and arranging them in a specified way. For example, a machine may assemble steel springs inside computer mice.

PROJECT MANUFACTURING

Project manufacturing or engineer-to-order (ETO) manufacturing is known and practiced in the industry but not in a formal way. Project manufacturing is to produce or assemble one unit of each unique product. Although it is a manufacturing environment, it follows the definition of the project of being temporary and unique. In mass production, there is a production or assembly line that produces thousands of units from a certain product. The methods used for production planning, scheduling, and controls of mass production cannot be employed to project manufacturing. Instead, a company that owns or operates manufacturing plants dedicated to manufacturing or assembly of projects use job-shop techniques to schedule and controls their production. Not only that, they try to implement state of the art production process improvements that are designed mainly for production of high volume such as mass production or job-shop manufacturing.

Blevins introduced the topic of project manufacturing in a very nice and simple way. He stated that the project manufacturing business has a set of islands and urged that they need to be integrated for a better planning and controls. Fox introduced a comprehensive list of challenges and sources of complexity that face project manufacturing. Some of these challenges are:

- Giving an authority to the customer. That is typical in project environments but not very helpful in scheduling project manufacturing.

- Change of priorities of individual customers makes the project manufacturing schedule to stop and resume several times during the lifecycle of the project.

- High number of components that are needed for the assembly of a single sub-product. For example a side wall of a boiler may need more than 100 tubes in different shapes and sizes.

- Components for the same sub-assembly may have a high variation in delivery times.

Abdelmaguid and Nassef stated that the job shop scheduling problem (JSP) is a traditional decision making problem that is encountered in low volume–high variety manufacturing systems which are known as job shops. Although job-shop scheduling is dedicated for low volume manufacturing, it is still a good technique for repetitive products but not as good for a project manufacturing. Every final product and all of its components are totally different from any other product. Therefore, it is unwise to apply an approach for repetitive products into non- repetitive parts.

Another simple reason to examine project manufacturing problems is that manufacturing the product is one part of the whole project. Integrating production schedule with the rest of the project schedule will be a nightmare if production is scheduled using job-shop approach. Caron and Fiore urged to find an innovative approach to integrate manufacturing and logistics with project management. Their approach was suitable back then before the popularity of the enterprise project management tools. On their study of construction project complexity, Bertelsenand Koskela urged that using two different systems for project management and project production adds to both the complexity of the project and the uncertainty of the objectives.

The following table shows some differences between the repetitive production and project manufacturing. These differences are given to emphasize the need of a different method for scheduling and controlling project manufacturing.

Category	Repetitive Manufacturing	ETO / Project Manufacturing
Products	Makes standard products	Products are unique
Pricing	Uses a price list	Estimates, quotes, and Bidding
Number of components per sub- assembly	Low	High
Inventory	Based on part number	No Inventory
Engineering	No or only a few engineering changes	A significant number of engineering changes
Value	Low value	Typically higher in value
Production routing	Standard	Customized for every product
Lead time	days or weeks	months or years
Shipping	Ships from finished goods	Ships from WIP(work-in-progress)
Progress measures	Measures cost variance from the standard cost.	Measures cost variance from the original budget.

The approach provides solutions to the two requirements of the project manufacturing business:

- Integrating the manufacturing schedule with the rest of the project schedule.

- Managing the overall load of the manufacturing plant using enterprise project controls systems.

In fact, this approach has been implemented in a few plants dedicated to project manufacturing of heavy power generation products. The approach simplified the planning and controls that used to be performed using job-shop methodologies. It provided more flexibility in stopping the project and resuming it again.

The proposed system can be integrated with the manufacturing execution system of the plant. That will make it fully automated which can provide real time status of the project. If there are not enough capabilities to integrate the system with manufacturing execution systems, it can be implemented by itself and the initial schedule and its progress could be updated manually.

Creation of a Manufacturing Project Schedule

For every project or product a unique routing sheet is always developed by the manufacturing engineering team. The routing sheet consists of a set of routing sheets for shippable products or subassemblies of the final product. Every shippable product is considered as a single node in the work breakdown structure (WBS) of the whole project. Each routing sheet at a WBS node consists of a set of work-orders. A work order describes the sequence of operations that are needed to produce a smaller subassembly. Products from work orders may be assembled using a set of operations described in another work order. A work order may have one or more operations.

Typically each operation in the routing sheet contains the following:

- Operation duration,

- Labor class,

- Number of needed man-hours,

- Work center,

- Number of machine hours,

- Material being processed,

- Number of material units.

In the critical path method (CPM) world, each operation can be considered an activity. Therefore, the routing sheet allows the creation of the activity list complete with its duration, labor resources, machine resources, and material resources. If the routing sheet is in electronic format, which is most probably the case, it can be easily converted into a list of activities in the project scheduling software. That activity list will be complete with all resources loaded. Moving activity data from the routing sheet into the scheduling system can be automated and can be performed in few moments even if the project is very large.

To have a complete schedule, one important feature is still missing. That is the activity sequence. Examining the workflow of project manufacturing, it found that for every project or product, there is a unique engineering bill of material (EBOM) not a standard BOM. Knowing that each work-order represents a component or part in the product, the EBOM specifies the parent assembly of each subassembly. That kind of parent-child relationships, allows for generating activity sequencing. Consider the following example.

Part C is a subassembly of part D. Part C is assembled by welding parts A and B. Part A is a 40 feet tube. It needs to 4 operations; cleaning, milling, bending, and heat treatment. Part B is a flange and needs only two operations; drilling and grinding. Then the sequence will be A1, A2, A3, A4, C1, C2, C3, D1. In parallel, there will be another sequence B1, B2, C1, C2, C3, D1 as shown in figure.

Following that procedure, activity sequencing can be easily generated and automated. The WBS of the project should be recognized and followed during generating the activity list either manually or automatically. Imposing activity sequence on top of that will produce a rotated shape of the EBOM.

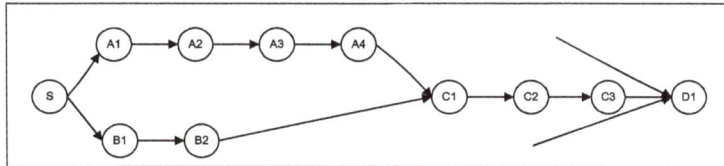

Activity sequencing example.

These two steps generate the initial resource loaded manufacturing schedule of the project. It is generated using the CPM and can be easily integrated with the rest of the project schedule. The benefit of the integration is known to most people in the field of project scheduling. Any changes in the engineering schedule will affect the manufacturing schedule as soon as they are recorded. Although such a situation is not favorable in a manufacturing environment, it is the fact of life. Changes on engineering schedules and their reflections on manufacturing schedules are much easier to handle using the integrated schedule than having two different scheduling systems. However, these changes complicate the resource management of the manufacturing plant as shown below.

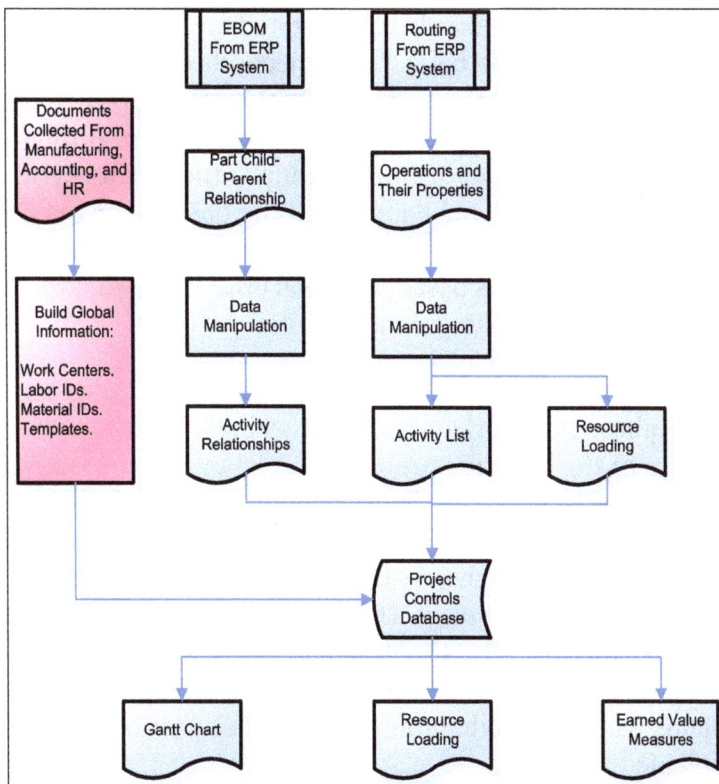

Process Data Flow.

Figure shows the flow of information to perform the procedure mentioned above. It is assumed that both the routing sheets and the EBOM are generated and posted into the ERP system of the performing organization. The two blocks in different color are assumed to be performed once in the lifetime of the system and then updated when changes happen. These are meant to build the global structures such organization chart and resource hierarchy of the manufacturing plant in the scheduling system.

Updating the Manufacturing Schedule

In most manufacturing plants, there is a manufacturing execution system which collects actual information from the shop floor. The system collects real-time information about each operation in the routing sheet. It collects the following information:

- Operation start time and date.

- Time elapsed on an operation.

- Labor hours charged to an operation.

- Machine hours charged to an operation.

- Material consumed so far.

- Number of units completed.

- Operation finish time.

A baseline based on the initial schedule should be generated for future comparison with the project progress.On a daily, weekly, or biweekly basis, the project schedule can be updated automatically by extracting the actual start date, actual duration, actual labor hours, actual machine hours, and actual material processed from the manufacturing execution system. If the process is automated, the update cycle might be daily. If it is performed manually the update cycle might me weekly or biweekly. The number of unites completed determines the physical or realistic percent complete of each activity. The earned value measures (EVM) can be employed with a great confidence since the percentage complete of the activity is based on a total objective measure. Therefore, the system will be able to report the project progress and more.

Suppose that the baseline states that the activity duration is 100 hours and it requires 200 labor-hours to bend 1000 tubes. Two workers will be working on this activity for 100 hours. After 5 working shifts, 40 hours of the duration are gone, 80 labor hours are charged, and 300 tubes are completed. Using the EVM, this activity is over budget and behind schedule. However, there is another conclusion that can be captured. It was assumed that each tube will be bent using 0.2 labor hours. The progress shows that each tube consumed 0.27 of labor hours. That concludes that there is a problem with the process efficiency. This is different from accelerating a construction or engineering activity to finish on time. The situation here is about the efficiency of the manufacturing process which should be corrected. The measures do not only give progress of the activity but also they report problems with the production process. Manufacturing management should pay attention to these problems.

Resource Loading for Project Manufacturing Plant

One of the biggest advantages of the enterprise control systems is the combination of resource requirements for different projects. Combining the demand on resources helps the manufacturing managers plan their resources and prioritize the workload. Figure shows a requirement for milling mechanist role for all projects currently active in the enterprise database (that are the active projects in the plant). This helps the manager of the milling group prepare for the peak resource allocation on milling mechanist by training other workers. Those with a secondary role as milling mechanist will serve as primary worker when they are needed.

Another planning scenario for the manager of the milling department is when he is negotiating with the project managers to move some of the activities back and forth in order to get rid of the demand peak. Of course the project manager needs to consult with the manufacturing scheduler on the possible effects moving an activity will cause on the overall schedule due to the dependency of other activities.

Is it possible to find one crew so heavily over-allocated? Yes, it is. Project managers usually know their employees. They know who gets the job right from the first time, or can give the initial results fast, or works with minimal supervision, or has many years of experience. During the initial planning of the project, they ask to put that crew on their project team. The result an over-allocation of a certain crew while under-allocation of other crews. It is the responsibility of the functional manager to reallocate the extra load to other crews who can also perform the job. The functional manager can coach the less experienced resource to achieve the job efficiently. These resources work for the functional manager who is assumed to know most of the details of their work. By keeping a close eye on the load of each resource in the group, the resource manager can avoid the situation of over-allocation from happening.

Changes on the start date of the project or the start and finish of certain activities are imposed by the client or the engineering group. These changes will force the project and its resource load to move back and forth. That will make the requirements on resources dynamic and uncertain as noted by Wullink Although it complicates the resource management process, it reflects the real life. The proposed system of managing resources using the CPM may seem to oversimplify the situation but implementing it in a few plants proved more efficient than computational systems. The CPM gives the true picture to the resource managers and leaves the decisions to them. Computational methods make the decisions without human involvements. They are also data hungry

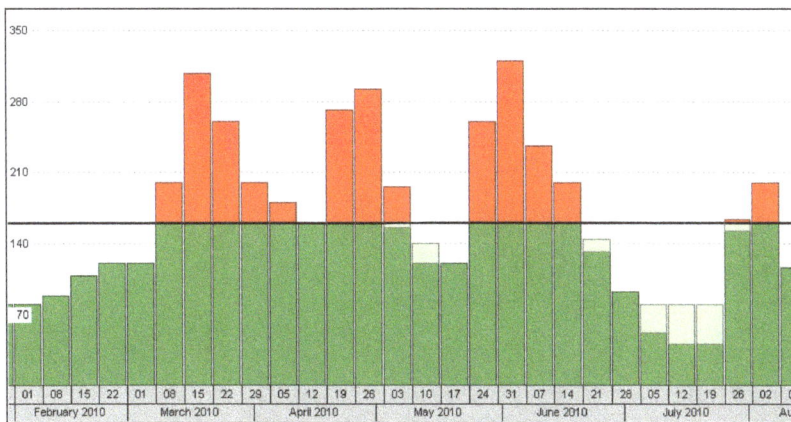

Role-Requirements Histogram.

Managing all resources needed for all active projects in the manufacturing plant helps in identifying bottlenecks. Manufacturing projects use high value machines like an overhead crane, a mobile 100 ton crane, a large sized milling machine, etc. If such a machine is overloaded, it represents a bottleneck and everything else will be scheduled around it. To upgrade from one unit of that equipment, another unit should be added. Because adding one unit is so expensive and requires a big investment, it is important to make sure that both the added unit and the existing units are almost fully utilized. In a project manufacturing plant, most resources are always under loaded with the exception of one work center, which is always over loaded. A typical project uses only some of the work centers. However, one work center is being used by all projects that come to the plant. This is the over allocated machine. When a new project is assigned to the plant, it is scheduled based on the availability of the over allocated work center.

The author was helping the plant by implementing the concepts and tools of enterprise project manufacturing systems. The project/production controls team was used to allocating only human resources, but was advised to allocate all resources needed for all activities. They did so for a few months until they found that the over allocated work center was the bottleneck driving the schedule of everything else. Actually, this bottleneck causes more problems to the plant than the scheduling problems do. In a follow up visit to the plant, it was found that they only allocated and planned for the bottleneck work center, while not paying any attention to the rest of the resources. Although it sounds reasonable, it is not recommended to forget about the wasted hours of the under allocated resources.

To maximize the utilization of the under allocated resources, the plant is advised to acquire a second unit of work center similar to the one that is over allocated. However, with that investment, will the second machine be reasonably utilized? There is a need to optimize between the two factors to make the decision that is best for the organization. If there are many projects assigned to the plant, in order for the two units to be reasonably loaded, then the plant would go with the option of acquiring another unit. If this assumption is not true, they may study another solution to minimize the waste in the under allocated areas.

The proposed system provides a methodology to answering all of these questions and to generating different what-if scenarios. It provides a the big picture of the resource loading for the plant in a dynamic environment that can be apply the optimization methods to level resources or leave the task to the resource managers as described above. It also provides a tool to respond to customers about estimates and quotes for new projects. Knowing the current and projected resource load, plant management can determine when a new job will fit in the plant schedule. Although the estimate is not a clear answer as noted by Wullink every customer requires a time and cost quotation for their project.

LEAN MANUFACTURING

The term Lean manufacturing refers to the application of Lean practices, principles, and tools to the manufacture of prototypes and products. Manufacturers are using Lean manufacturing principles to eliminate waste, optimize processes, cut costs, and boost innovation in a volatile market. Lean manufacturing defines waste as anything that doesn't add value to the customer. This can be a process, activity, product, or service anything that requires an investment of time, money, and talent that does not create value for the customer is waste. Lean manufacturing provides a systematic method for minimizing waste within a manufacturing system, while staying within certain margins of control such as productivity and quality.

Many early Lean practitioners focused on the tools and methods required to improve. Examples of Lean production methods in use today include SMED (Single Minute Exchange of Die), TPM (Total Productive Maintenance) and Kanban, while examples of methodologies used in Lean Manufacturing include Lean Six Sigma and DMAIC (Define, Measure, Analyze, Improve and Control). Each of those movements has its own unique merits, but it would be fair to say that the style of management they were aiming for is now generally thought of as Lean.

While you might not think of them as specifically lean manufacturing principles, you may recognize some of these concepts:

- Value stream mapping.

- Demand-based (pull) system.

- Continuous improvement.

- Measurement, KPIs, and Visualization.

Key Lean Manufacturing Principles

In order to remain competitive, manufacturing organizations have to be in a constant mode of continuous improvement - finding ways to improve time to market (TTM) and increase innovation, while lowering costs, streamlining processes, and adapting new technologies.

Lean manufacturing offers a holistic approach to improvement, with the methods, tools, and cultural ideals that companies need to stay innovative and agile. These Lean manufacturing principles enable organizations to become nimbler and more innovative.

Value Stream Mapping

Value stream mapping can be used to improve any process where there are repeatable steps – and especially when there are multiple handoffs. Much of the waste in knowledge work occurs in the handoffs (or wait time) between team members, not within the steps themselves. Inefficient handoffs in knowledge work may not look like bottlenecks on a car assembly line, but they produce the same effect: decreased productivity, overwhelmed workers, and lower work quality.

Analyzing and aiming to eliminate these inefficiencies at the organizational level is the first step toward becoming leaner. Lean actions can be focused on specific logistical processes, or cover the entire supply chain.

For example, an analysis of a SKU would look like this: First you would aim to visualize its path, evaluating all the participants from material suppliers to the consumer, and then conduct a gap analysis to determine necessary next steps to improve the value stream and achieve the objective.

You'd then make these small improvements over time throughout the supply chain, increasing organizational learning and streamlining the process of creating that SKU.

Demand-based Flow (Pull) Manufacturing

In a pull manufacturing system, inventory is only pulled through each production center when it is needed to meet a customer's order. Pull systems allow "just-in-time" delivery of work. Unlike other work methods that allow for an unlimited amount of work at once, a pull system enables everyone at a specific organizational level to focus on one thing (or just a few things) at one time.

Benefits of using a Kanban control system or pull system include:

- Ability to manage change.

- Ability to quickly adapt work to new information.

- Increased ability to scale the team to the appropriate size for the project.

As they work through a list of "to-do" items in a backlog, team members pull new tasks only as old tasks are completed. This way, when something changes that impacts the business requirements (as it always does), the team can quickly adapt, knowing that the majority of work they have already completed can still be applied to the project. Finally, because teams using a pull system are self-managed to a certain degree, pull systems contribute to the scalability of a team, or the ability for a team to accommodate different sized projects while remaining cohesive.

For manufacturers, this means teams can be more agile, deliver faster, and innovate faster and more strategically. Organizations who adopt a Lean pull system are also able to significantly improve the reliability and accuracy of forecasting for their suppliers and customers.

Continuous Improvement Mindset

An organization-wide commitment to continuous improvement is essential for sustainable success with Lean manufacturing. At its core, Lean is continuous improvement it's improving product and process while eliminating redundant, excessive, or inefficient activities. Continuous improvement can be viewed as a formal practice or an informal set of guidelines but it must be well integrated into the culture of an organization in order to make a meaningful and lasting difference.

Measurement, KPIs and Visualization

Lean manufacturing metrics, such as lead time, cycle time, throughput, and cumulative flow, help organizations measure the impact of their improvement efforts. Collecting, analyzing, visualizing, and socializing these metrics (through shared dashboards) is essential to promoting transparency and driving change.

Successful lean manufacturers use up-to-date dashboards at the team, leader, and executive levels to paint an accurate picture of the impact that changed processes are having. It should be noted that the emphasis is on surfacing key performance indicators of processes not people. This reinforces a collective responsibility by teams to pursue opportunities for improvement and focus on value creation for customers.

AGILE MANUFACTURING

Agile manufacturing is a modern approach or strategy used by manufacturers to respond quickly to the changing customers' needs and market demands. There are multiple factors that enable you to become an agile manufacturer – modular and customer-focused product design, Information Technology, corporate partners and knowledge culture.

Agile manufacturing enables the organization to respond to the market demands in a jiffy, and that too without compromising on the quality. Consumers appreciate the speed and so do your business. Without compromising on the quality of products, an agile manufacturing plan guarantees fast delivery services and promptness. There has to be an intuitive agile manufacturing plan which should focus on the rapid moment for creating and executing idea while gauging what is right for the business and what isn't.

How is Agile Manufacturing Different from other Practices?

Modular, Customer-focused Product Design

Agile manufacturing plan encourages quicker modification and adaptability. When products are designed in the modular pattern, variations are done easily. It brings together different pieces of materials to simplify changing designs and deliver products with an unprecedented level of speed and personalization.

Information Technology

Organizations, whether large scale or SMBs aim to improve external and internal business communication. Proper IT infrastructure drives agile manufacturing and ensures that employees are updated with every managerial decision and technological advancement.

Corporate Partners

Not every business model aims to build good relationships with the partner companies. Agile manufacturing plans lay emphasis on maintaining short-term partnerships and co-operative projects. However, it is always suggested to examine your corporate relationships prior to implementing any new plan.

Knowledge Culture

Training is the key to agile manufacturing. It ensures that the right amount of knowledge is cascaded to the suitable person because everyone is the organization should seamlessly understand the rapid changes and adaptation through which the business is going. Appropriate training should be evenly provided to all the concerned people in order to succeed in agile manufacturing.

Lean Manufacturing

Agile manufacturing adds an extra dimension to lean manufacturing. Talking about lean manufacturing, it is a systematic approach of minimizing waste while maintaining the quality and production level within a manufacturing unit. On the other hand, agile manufacturing is more focused on fulfilling the customers' demand in minimum time. This method is the best suitable for businesses that undergo drastic transformation and fluctuating customer needs. Considered as a precursor to agile, lean manufacturing has paved the road to modernization.

It aims to create a structured, well-organized and encouraging work culture while ensuring minimum wastage. The relationship between lean and agile manufacturing can be simply explained with a simple example – one is a lean or thin person and the second is a fit one. Both are obviously not the same. Here, the latter one is an agile manufacturer and the former one is a lean manufacturer.

Both are not the same, however, a company can be lean or an agile plan driven. And when the nuances of both lean and agile manufacturing are inculcated in an organization, it is known as leagile (lean + agile) manufacturing.

In a lean and agile process, emphasis is on not only quantitative inputs, but also qualitative comments, assumptions, risks, and opportunities associated with the plan. Audit trail capability is used for all the stakeholders' inputs to add transparency and accountability to the process and encourage open and honest communication. This also increases participation and helps create a single version of the truth with no hedging, underestimating, or overestimating.

In lean and agile manufacturing, there is continuous horizontal reconciliation among stakeholders and vertical integration among strategic, tactical, and operational planning, which are interwoven harmoniously in order to create a cohesive fabric. In addition to being lean, this tactic accentuates inventory and capacity flexibility and cushions against market volatility and forecast inaccuracies, strengthening the organization. Capacity and inventory buffers act as the shock absorbers between strategic planning and operational planning in the event of unpredictable events. What-if scenarios for alternative courses of action and contingency plans help the organization mitigate risk and enable greater responsiveness to changing customer and market needs.

How can ERP Drive Agile Manufacturing?

In today's era, agility is not just a hallmark but also a requirement for surviving in this competitive marketplace. And to drive manufacturing agility, modern Enterprise Resource Planning solution encourages lean manufacturing capabilities and just-in-time production to handle growing demands or sudden market changes. Enterprise Resource Planning software is the key to agile manufacturing. OptiProERP with SAP Business is a reliable suite of business management tools that offer advantages such as effectiveness, accuracy, and scalability to meet your business requirement.

It inculcates a huge array of advanced features which promotes capacity planning, material planning, quality control, production, inventory management, purchase, sales, accounts, finance and more.

Principles within Agile Manufacturing

Agile manufacturing can aid manufacturing operations by focusing on personalized customer products. Before applying agile methodology to the operation, it is important to analyze the principles within agile manufacturing: consumer enrichment, competitive enhancement, organization, and leveraging impact.

Consumer Enrichment

While lean is more waste oriented, agile is more customer oriented. One of the most important principles within agile is enriching the customer through various factors such as identification, monitoring, and understanding factors such as Quality Function Deployment. Satisfying consumer demands is a key component within agile manufacturing.

Competitive Enhancement

Having all departments on board for agile methodology can ensure for a much more efficient and competitive atmosphere. This is by partnering with firms that have the same ideas and mindset about the production. This is how you can set theself a step above competitors and adopt a much more flexible and adaptable supply chain.

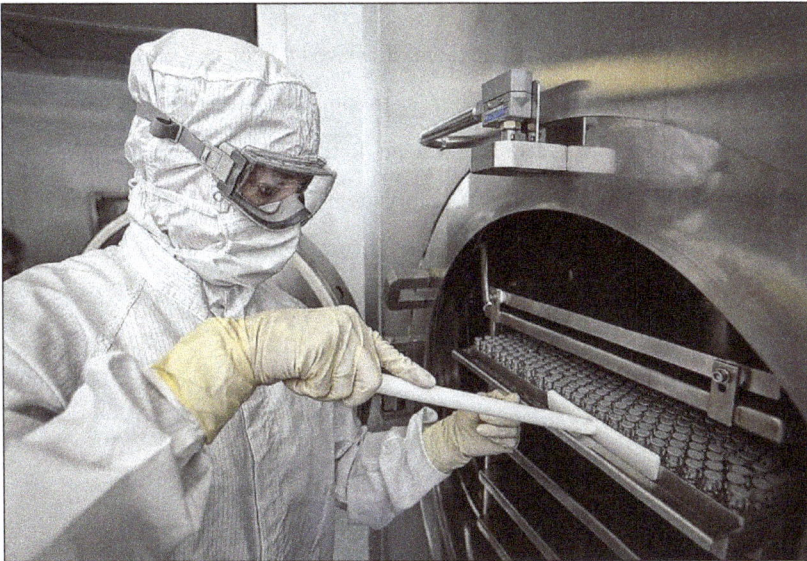

Organization

Proper organization within the operation is one of the most important aspect of an agile manufacturing operation. This is due to swift changes in circumstances such as consumer preference, demand, and production. This allows production to be flexible and be prepared for a change at a moment's notice.

Leveraging Impact

People are essential within agile operations, which is why it is important to constantly monitor the impact of human capital. This is because humans possess skill, information, and the drive to enhance productivity and improve the manufacturing process. Locating potential leaders that can take production in the right direction can bring extreme benefit to an agile operation. It is also extremely important to keep up with current manufacturing trends and advancements in technology, which can improve the manufacturing operation tremendously.

Advanced Planning and Scheduling Software (APS)

Advanced planning and scheduling software (APS) can enhance agile manufacturing operations with ease. As the software is able to be easily integrated with ERP or MRP operations, it offers various benefits and capabilities that can optimize production tremendously. Various benefits of the software include the following:

- Improved Delivery Performance.

- Profit Boosts.

- Reduction in Inventory and Cost.

- Six Month ROI.

References

- Definition-manufacturing-industry: bizfluent.com, Retrieved 5 February, 2019

- What-is-discrete-manufacturing, glossary, en-us: sage.com, Retrieved 6 March, 2019

- What-is-lean-manufacturing, lean, learn: leankit.com, Retrieved 7 April, 2019

- What-is-agile-manufacturing-and-how-can-it-help-you-succeed, in: optiproerp.com, Retrieved 8 May, 2019

- Principles-within-agile-manufacturing: planettogether.com, Retrieved 9 June, 2019

Manufacturing Engineering

The branch of engineering which focuses on developing tools, processes, machines, etc. and integrating them to maximize the revenue is called manufacturing engineering. It is mainly bifurcated into manufacturing production and packaging engineering. This chapter discusses about manufacturing engineering and its aspects in detail.

Manufacturing engineering is that branch of professional engineering requiring such education and experience as is necessary to understand and apply engineering procedures in manufacturing processes and methods of production of industrial products. It requires the ability to plan the practices of manufacturing; to research and develop tools, processes, machines, and equipment; and to integrate the facilities and systems for producing quality products with the optimal expenditure of capital.

Manufacturing engineering embraces the activities in the planning and selection of the methods of manufacturing, development of the production equipment, and research and development to improve the efficiency of established manufacturing techniques and the development of new ones. These' activities include but are not limited to:

- Facilities planning, including processes, plant layout, and equipment layouts.

- Tool and equipment selection, design, and development.

- Value analysis and cost control relating to manufacturing methods and procedures.

- Feasibility studies for the manufacture of new or different products with respect to the possible integration of new items into existing facilities.

- Review of product plans and specifications, and the possible changes to provide for more efficient production.

- Research and development of new manufacturing methods, techniques, tools, and equipment to improve product quality and reduce manufacturing costs.

- Coordination and control of production within a plant and between separate plants.

- Maintenance of production control to assure compliance with scheduling.

- Recognition of current and potential problems and the implementation of corrective action to eliminate them.

- Economic studies related to the feasibility of acquiring new machinery, tools, and equipment.

Manufacturing engineering is divided into four basic functional areas:

- Manufacturing planning is the preliminary engineering work relative to the establishment of a manufacturing system for the production of a product. It includes the selection and specification of the necessary facilities, equipment, and tools, as well as the plant layout to provide the most efficient operation.

- Manufacturing operations is the engineering work involved in the routine functioning of an existing plant or facility to provide efficient and economical production output to quality standards. It includes the improvement of exist-ing layouts, procedures, tooling, and product plans and specifications.

- Manufacturing research is the pursuit of new and better materials, methods, tools, techniques, and procedures to improve manufacturing processes and reduce costs. It includes the creation of concepts and innovative uses of existing items.

- Manufacturing control is the management of manufacturing operations to assure compliance to required schedules. It includes coordination of all manfacturing departments and support departments, such as purchasing and materials.

This handbook has been prepared with the goal of setting standard practices within the area of modern manufacturing. In order to present the most current practices, many commercial, trade-association, and military requirements have been taken into consideration. When the practices presented herein are applied properly, they will resolve most situations that arise in everyday routines.

Since the proper application of the practices presented can only be indicated in a general way, you must always ask yourself, Who will use it and for what?, before useful work can begin. Then, once started in the right direction, these practices can be useful as tools to accomplish the desired result-that is, a com-pleted product that is accurate, complete, and readily usable as intended.

This handbook has not been prepared as a textbook for beginners. Neverthe-less, the chapters are arranged in progressive order to guide new engineering personnel into accepted practices. It also provides a convenient reference for all other personnel who work in and with the manufacturing department.

MANUFACTURING PRODUCTION

Manufacturing production refers to the methodology of how to most efficiently manufacture and produce goods for sale, beyond just a bill of materials. Three common types of manufacturing production processes are make to stock (MTS), make to order (MTO) and make to assemble (MTA). Such strategies have advantages and disadvantages in labor costs, inventory control, overhead, customization, and the speed of production and filling orders.

Manufacturing is the creation and assembly of components and finished products for sale on a large scale. It can utilize a number of methods, including human and machine labor, and biological and chemical processes, to turn raw materials into finished goods by using tools. Production is similar but broader: It refers to the processes and techniques that are used to convert raw materials or semi-finished goods into finished products or services with or without the use of machinery. Whether it is one or the other, manufacturers need to match their production methods to the needs and desires of the market, the available resources, order volume and size, seasonal shifts in demand, overhead costs (such as labor and inventory), and numerous other variables.

Make to Stock

The make to stock strategy is a traditional production strategy that is based on demand forecasts. It is best utilized when there is predictable demand for a product, such as for toys and apparel at Christmastime. MTS can be problematic when demand it more difficult to predict, however. When used with a business or product that has an unpredictable business cycle, MTS can lead to too much inventory and a dent in profits, or too little and a missed opportunity.

Make to Order

The make to order strategy (also known as "built to order") allows customers to order products built to their specifications, which especially useful with heavily customized products like computers and computer products, automobiles, heavy equipment, and other big-ticket items. Companies can alleviate inventory problems with MTO, but customer wait time is usually significantly longer. This demand-based strategy cannot be used with all product types.

Make to Assemble

The make to assemble strategy is a hybrid of MTS and MTA in that companies stock basic parts based on demand predictions, but do not assemble them until customers place their order. The advantage of such a strategy is that it allows fast customization of products based on customer demand. As such, a good example is found in the restaurant industry, which prepares a number of raw materials in advance and then awaits a customer order to start assembly. One downside to MTA is if a company receives too many orders to handle with the labor and components it has on hand.

Make to Order (MTO)

Make to order (MTO), or made to order, is a business production strategy that typically allows consumers to purchase products that are customized to their specifications. It is a manufacturing process in which the production of an item begins only after a confirmed customer order is received. It is also known as mass customization.

This type of manufacturing strategy is referred to as a pull-type supply chain operation because products are only made when there is firm customer demand. The pull-type production model is employed by the assembly industry where the quantity needed to be produced per product specification is one or only a few. This includes specialized industries such as construction, aircraft and vessel production, bridges, and so on. MTO is also appropriate for highly configured products such as computer servers, automobiles, bicycles, or products that are very expensive to keep inventory.

How Make to Order Strategies Work

The make to order (MTO) strategy means that a firm only manufactures the end product once the customer places the order, creating additional wait time for the consumer to receive the product, but allowing for more flexible customization compared to purchasing directly from retailers' shelves.

In order to manage inventory levels and provide an increased level of customization, some companies adopted the make to order production system. The MTO strategy relieves the problems of excess inventory that is common with the traditional make to stock strategy. Dell Computers is an example of a business that uses the MTO production strategy, wherein customers can order a fully customized computer online and receive it in a couple of weeks.

The main advantage of the MTO system is the ability to fulfill an order with the exact product specification required by the customer. Sales discounts and finished goods inventory are also reduced, and stock obsolescence is managed. However, for an MTO system to succeed, it should be coupled with proactive demand management. It should also be considered that the MTO system is not appropriate for all types of products.

Related to MTO is assemble to order (ATO), which is a business production strategy where products ordered by customers are produced quickly and are customizable to a certain extent. The assemble-to-order (ATO) strategy requires that the basic parts of the product are already manufactured but not yet assembled. Once an order is received, the parts are assembled quickly and sent to the customer.

Make to Order versus Make to Stock

Traditional production methodologies produce products and stock them as inventory until a customer buys them. This is known as make to stock or MTS. However, this system may be prone to wastage and obsolescence, as inventory sits on shelves awaiting purchase. This problem is particularly acute in an industry like technology, where the pace of advancement is quick and the problem of obsolete inventory could quickly arise.

In theory, the MTS method is a great way for a company to prepare for increases and decreases in demand. However, inventory numbers and, therefore, production, are derived by creating future demand forecasts based on past data.

There is a high likelihood that the forecasts will be off, even if by just slightly, meaning that a company might be stuck with too much inventory and too little liquidity. This is the main drawback to the MTS method of production. Inaccurate forecasts will lead to losses, stemming from excess inventory or stockouts, and in fast-paced sectors such as electronics or computer tech, excess inventory can quickly become obsolete.

Limitations of Make to Order

The two main drawbacks of make to order management are timeliness and cost of customization. If products are already on the self as with MTS, then a customer need not wait until the product is made, assembled and delivered to spec. Cost is also a factor; pre-made and available products are all alike and so manufacturing costs are lowered due to economies of scale. Made to order will tend to be more expensive for the consumer since it involves customizable parts and finishes.

Make to assemble (MTA)

A make to assemble or MTA strategy is a manufacturing production strategy wherein a company stocks the basic components of a product based on demand forecasts but does not assemble them until the customer places an order. This allows for order customization. MTA production is basically a hybrid of two other major types of manufacturing production strategies: make to stock (MTS) and make to order (MTO).

With MTS, businesses base their production on demand forecasts and final products are assembled before customers have ordered them. Customers can thus get items quickly, but only if the correct quantities have been manufactured, and businesses risk overproduction. At the opposite end of the spectrum, MTO creates items to customer specifications after they are ordered, so it is sometimes a slow process. The MTA production strategy is not as flexible for businesses as the MTO strategy, though MTA allows customers to get their orders sooner.

Industries that use Make to Assemble

Aspects of the foodservice and restaurant industry may use a make to assemble strategy when serving customers. In a restaurant, the ingredients for an entrée may be present in the establishment's refrigerator, awaiting assembly when a customer requests the item. The degree of assembly may vary, as certain parts of the dish might be premade or precooked. For instance, a quick-serve restaurant may use some frozen food items that simply need to be heated before being added to other items that are part of the order.

The make to assemble strategy may be adopted by independent fabricators and product makers who sell their wares through marketplaces while maintaining unassembled parts in storage. For example, sellers on platforms such as Etsy might stock the pieces to create apparel or accessories they offer for sale. Similar sellers who use 3D printers might adopt a comparable strategy, keeping parts and pieces they crafted at the ready to be assembled when a customer puts in an order.

Reasons for using Make to Assemble

Reasons for using a make to assemble strategy for production vary, though may be based on ease of storage or the shelf life of the product. Food products, for example, typically have a window of time when the items remain fresh. The final product, once complete, might also have a short time when it is edible. Storing the ingredients separately until they are needed is a common way to be more efficient. Depending on the type of product, it may be logistically more feasible to store the parts rather than the final product.

Make to Stock (MTS)

In MTS (Make to Stock), products are manufactured based on demand forecasts. Since accuracy of the forecasts will prevent excess inventory and opportunity loss due to stockout, the issue here is how to forecast demands accurately.

MTS (Make to Stock) literally means to manufacture products for stock based on demand forecasts, which can be regarded as push-type production. MTS has been required to prevent opportunity loss due to stockout and minimize excess inventory using accurate forecasts. In the industrialized society of mass production and mass marketing, this forecast mass production urged standardization and efficient business management such as cost reduction.

As an economy expands, the income of consumers increases and so demand also continuously increases. Demand changes according to the boom and bust cycle of the economy. Even if demand decreases and inventory increases, inventory will turn into cash one day when demand recovers. Therefore, the main theme of business management is how to predict the future based on the demand fluctuation cycle of the past. In specific, the development of a production/inventory management system is needed to improve management efficiency by, for example, setting safety stock, optimal production, and ordering points based on lead times of material procurement, production, and delivery as well as demand forecasts.

If demand can be accurately forecasted to some extent then there is no problem in creating a forecast production schedule. If MTO (Make To Order) is like an elevator because MTO starts by receiving an order as an elevator starts by pressing a button, MTS (Make to Stock) is like a train schedule (supply schedule) for which the number of passengers (forecast demand) for each time period can be prospected from the past data. Most of daily necessities such as processed foods, sundries, and textiles are MTS-type products and quick response to consumers' needs (i.e. filling retailer's inventory) will minimize opportunity loss.

One issue of MTS is to handle supply management so as not to have excess inventory. Therefore, small-batch supply should be frequently performed by pull-type demand such as QR (quick response), ECR (efficient consumer response), CRP (continuous replenishment program), and VMI (vendor managed inventory). By doing so, product flow will accelerate and cash flow will increase. Changing push-type MTS to pull-type supply chain models such as CRP and VMI is the key to successful supply chain management.

PACKAGING ENGINEERING

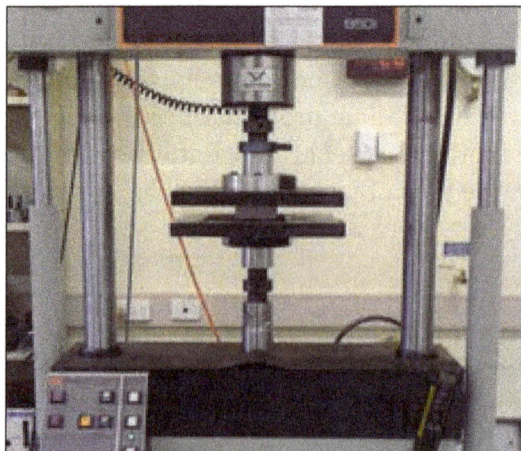

Packaging engineering is focused on the development and optimisation of protective packaging systems. Protective packaging systems are commonly used for products such as fruit and vegetables, beverages, mobile phones, toys, etc. However, it is also used in other contexts, such as protecting equipment or even people from damage or discomfort. Packaging engineering is interdisciplinary because it draws from many disciplines, mostly from mechanical (materials, dynamics) and chemical engineering (materials, food spoilage).

For instance, those who specialise in chemical engineering can focus on packaging systems to protect food from spoiling. On the other hand, the mechanical engineering components are generally focused on the dynamics of packaging, and how they protect products and equipment from damage due to vibrations, heat, etc. A packaging engineer should be able to determine a product's fragility, and have a good understanding of the supply chain and any environmental situations which can cause damage or spoilage.

Packaging engineers are also responsible for designing and optimising protective packaging systems, and therefore sustainability is a key factor in this discipline. Packaging can also be focused on its interaction with consumers, and how a product is packaged can have a significant influence on its profitability. You will commonly find packaging engineers working in a wide range of different fields, such as food and beverage (e.g. Coca Cola, Tesco, etc.), consumer products, electronics (e.g. smartphones), and even in the automotive and aerospace sectors (e.g. Tesla).

References

- Manufacturing-production, terms: investopedia.com, Retrieved 10 July, 2019

- Make-to-order, terms: investopedia.com, Retrieved 11 August, 2019

- Make-to-assemble, terms: investopedia.com, Retrieved 12 January, 2019

- Mts-make-to-stock, scm-terminology: lean-manufacturing-japan.com, Retrieved 13 February, 2019

- What-is-packaging-engineering, topic: whatisengineering.com, Retrieved 14 March, 2019

Understanding Manufacturing Processes

Manufacturing processes comprise of numerous procedures such as mesoscale manufacturing, laser engraving, casting, lost-foam casting, permanent mold casting, molding, blow molding, thermoforming, lamination, dip molding, rotational molding, thermal spraying, etc. This chapter has been carefully written to provide an easy understanding of these manufacturing processes.

Manufacturing process is basically a complex activity, concerned with people who've a broad number of disciplines and expertise and a wide range of machinery, tools, and equipment with numerous levels of automation, such as computers, robots, and other equipment. Manufacturing pursuits must be receptive to several needs and developments.

Beside above, technicians must understand the basic needs of workshop routines in terms of man, equipment, material, methods, revenue and other infrastructure conveniences needed to be placed properly for maximum shop or plant layouts and other support solutions effectively regulated or positioned in the field or industry within a properly planned manufacturing firm.

The complete knowledge of fundamental workshop technology and manufacturing processes is highly troublesome for anybody to claim competence over it. It deals with numerous aspects of workshops procedures also for providing the basic working awareness of the various engineering materials, tools, accessories, manufacturing processes, basic concepts of machine instruments, production criteria's, traits and uses of numerous testing instruments and calibrating or inspecting units for checking materials or products designed in various production shops in a commercial environment. It also explains and illustrates the use of several hand tools (calibrating, marking, forming and supporting gear etc.), tools, machinery and diverse methods of production that facilitate forming or shaping the existing raw materials into appropriate usable forms. Below are some of the manufacturing processes that are worth reading.

Types of Manufacturing Processes

Following are the 4 types of manufacturing processes.

Machining

Tools used for machining are immobile power-driven units used to form or shape solid materials, specifically metals. The forming is done by removing extra materials from a work-piece. Machine tools make up the foundation of advanced industry and are utilized either indirectly or directly in the manufacturing of tool parts.

They are categorized under three main categories:

- Traditional Chip-making tools.

- Presses.

- Modern machine tools.

Traditional chip-making tools form the work-piece by trimming away the unwanted part accessible as chips. Presses implement a several shaping processes, which includes shearing, pressing, or elongating. Non-traditional machine tools implement light, electric powered, chemical, and sonic power; superheated gas; and high-energy compound beams to form the exotic supplies and materials that have been created to meet the requirements of modern technology.

Joining

Every joining approach has particular design needs, while certain joint needs may propose a particular joining approach. Design for assembly, and fastener selection apply their own specifications.

Bolting is a standard fastening method, for instance, but welding may cut down the weight of assemblies. Naturally, joints intended for the two approaches would differ tremendously.

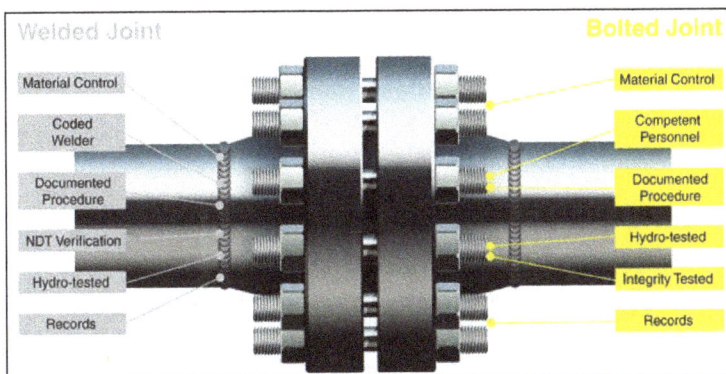

However, all joint patterns must consider features such as load factors, assembly effectiveness, operating surroundings, overhaul and upkeep, and the materials chosen.

Welding is generally a cost-effective approach to fabricate. It doesn't require overlapping materials, and so it removes excess weight brought on by other fastening methods. Fasteners don't have to be purchased and stored in stock. Welding also can minimize costs related to extra parts, for example angles mounted between parts.

Forming

Metal forming is the approach of creating the metallic components by deforming the metal but not by removing, cutting, shredding or breaking any part. Bending, spinning, drawing, and stretching are a few important metal forming process in manufacturing. The metal press such as die and punching tools are implemented for this manufacturing process.

Advantages: Same equipment can be utilized for manufacturing various components by simply changing the dies.

Disadvantages: High apparatus and tooling expenses. Heat treatment must be applied afterwards.

Casting

Casting is a manufacturing process in which a solid is dissolved into a liquid, heated to appropriate temperature (sometimes processed to change its chemical formula), and is then added into a mold or cavity. Thus, in just one step, complex or simple shapes can be crafted from any kind of metal that has the capability to be melted. The end product can have practically any arrangement the designer wants.

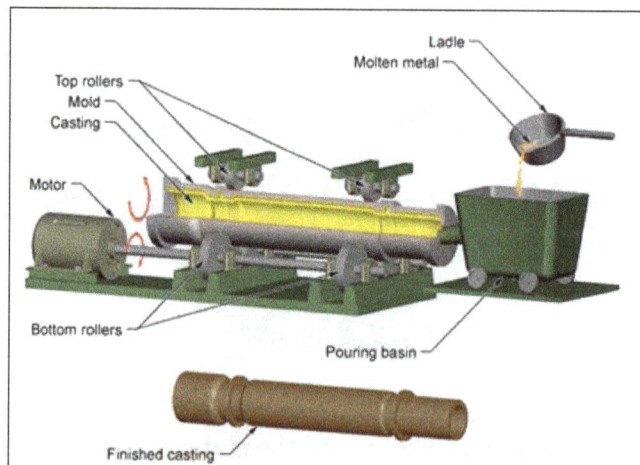

Furthermore, the reluctance to working challenges can be improved, directional attributes can be managed, and a pleasing look can be developed.

MESOSCALE MANUFACTURING

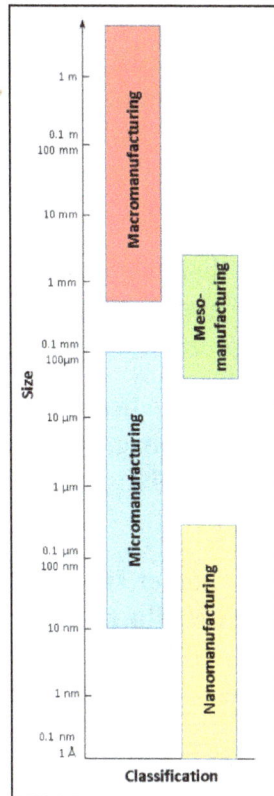

Manufacturing Classification.

Mesoscale manufacturing is the process of creating components and products in a range of approximately from 0.1mm to 5mm with high accuracy and precision using a wide variety of engineering materials. Mesomanufacturing processes are filling the gap between macro- and micromanufacturing processes and overlaps both of them. Other manufacturing technologies are nanoscale (< 100 nm), microscale (100 nm to 100 µm) and macroscale manufacturing (> 0.5 mm).

Applications

Application of mesomanufacturing include electronics, biotechnology, optics, medicine, avionics, communications, and other areas. Specific applications include mechanical watches, and extremely small motors and bearings; lenses for cameras and other micro parts for mobile telephones; micro-batteries, mesoscale fuel cells, microscale pumps, valves, and mixing devices for micro-chemical reactors; biomedical implants, microholes for fiber optics; medical devices such as stents and valves; mini nozzles for high-temperature jets; mesoscale molds; desktop- or micro-factories, and many others.

Processes

Manufacturing in the mesoscale can be accomplished by scaling down macroscale manufacturing processes or scaling up nanomanufacturing processes. Macroscale techniques like mill and lathe machining have been successful used to create features in the range of 25 μm. Meso Machine tools (mMTs), for example miniaturized milling machine, is a expansion of using traditional macroscale techniques to manufacture mesoscale products. With the limitation of self-excited vibration of machine tools and fatigue, microassembly and micro- and mesoscale milling are created to improve the maximum stiffness and dynamic operation of the milling process, which improves the overall performance of manufacturing. The development of mMTs has revealed many challenges that are specific to machining at the small scales. These challenges stem from the large influence of grain size at small scales and the necessity of extremely small tolerances for both the machine tools and the measuring tools.

Laser machining is a traditional technique that uses nanosecond pulses of ultraviolet light to create mesoscale features like holes, fillets, etc. The removal of material during laser machining is proportional to exposure time and therefore this process can be used to create three dimensional features.

A less traditional technique is to use focused ion beam sputtering (FIB) to remove material. This process involves focusing a beam of ions, like from gallium, to the work piece and this causes material to be removed. Using FIB sputtering has a relatively low rate of material removal and therefore has limited application.

Electrical discharge machining (EDM) is another subtractive manufacturing process used in the mesoscale. This process requires that electricity be transferred between the tool electrode and the work piece and therefore it can only be used to manufacture materials that conduct electricity. One advantage of EDM is that it can be used on hard materials that do not work well in traditional machining processes, such as titanium.

LASER ENGRAVING

Laser Engraving (or Laser Etching) is a Subtractive Manufacturing method, that uses a laser beam to change the surface of an object.

This process is mostly used to create images on the material, that may be seen at eye level. To do so, the laser creates high heat that will vaporize the matter, thus exposing cavities that will form the final image. It is using the laser for marking the surface of an item.

Working

This method is quick, as the material is removed with each pulse of the laser. The depth of the marks is controlled by the number of times the laser beam is passing on the material.

There are different types of laser engraving machines:

First ones are engraver machines where the workpiece stay stable, only the laser moves (or inversely). Other machines dedicated to cylindrical workpieces. The last possibilities are laser engraving machines where the laser and the workpiece are both immobile but galvo mirrors are moving the laser beam on the surface to engrave.

Uses

It can be used on almost any kind of metal, plastic, wood, leather or glass surface. You can get a lot of different engraved materials. Furthermore, it is more effective than traditional engraving for small objects, such as jewelry. There are also smaller chances to damage or deform the material. It can be used for many different applications, such as medical devices, fine art and so on.

It can be used for industrial applications.

Difference between Laser Etching and Laser Engraving

Laser Etching is a subset of Laser Engraving. In this particular case, the heat from the beam is not used to vaporize the matter, but to melt it. It is mostly used on metals, and will expand the material by creating a raised cavity. It thus modifies its characteristics, such as its reflectivity, and will create a contrast with its surroundings.

Compared to other processes using lasers, such as Laser Cutting, Laser Engraving keeps the initial shape of the material, and is the most common option to create personalized or customized objects. Logos, serial numbers, or pictures are typically the kind of creations thatcan be done thanks to Laser Engraving.

Laser Marking is less common but quite popular in the medical device industry. One of the major difference with Laser Engraving and Laser Etching is that here, the oxidation during the marking process make the material turn black. This laser marker technique is also good for logos or bar codes.

Laser Engraving: list of file formats

To use such a process, you need to provide a 2D file, such as a vector graphics or drawing of your image. The most commonly used formats are:

- .ai
- .bmp
- .cdr
- .dwg
- .dxf
- .eps
- .jpg/jpeg
- .pdf
- .png
- .scad
- .svg
- .xps

This Laser Engraving system does not work with any 3D file format, such as .stl, so be careful.

The model will then be converted into dots, and the distance between two of these dots will determine the depth of the engraving. Then the 2D model will be sent to an engraver machine.

The machines used for Laser Engraving or Laser Marking are the same that the ones used for Laser Cutting.

Casting

Molten metal before casting.

Casting iron in a sand mold.

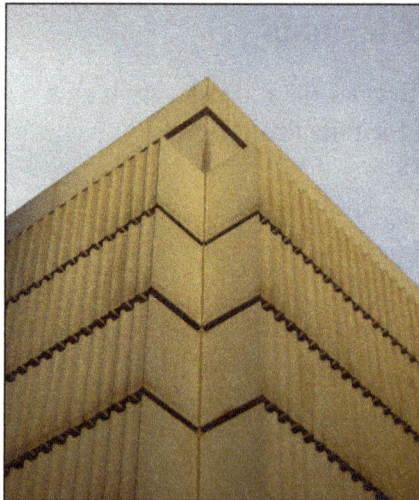

Judenplatz Holocaust Memorial (Nameless Library), by Rachel
Whiteread. Concrete cast of books on library shelves turned inside out.

Casting is a manufacturing process in which a liquid material is usually poured into a mold, which contains a hollow cavity of the desired shape, and then allowed to solidify. The solidified part is also known as a casting, which is ejected or broken out of the mold to complete the process. Casting materials are usually metals or various time setting materials that cure after mixing two or more components together; examples are epoxy, concrete, plaster and clay. Casting is most often used for making complex shapes that would be otherwise difficult or uneconomical to make by other methods. Heavy equipment like machine tool beds, ships' propellers, etc. can be cast easily in the required size, rather than fabricating by joining several small pieces. Casting is a 7,000-year-old process. The oldest surviving casting is a copper frog from 3200 BC.

Types

Metal

In metalworking, metal is heated until it becomes liquid and is then poured into a mold. The mold is a hollow cavity that includes the desired shape, but the mold also includes runners and risers that enable the metal to fill the mold. The mold and the metal are then cooled until the metal solidifies. The solidified part (the casting) is then recovered from the mold. Subsequent operations remove excess material caused by the casting process (such as the runners and risers).

Plaster, Concrete or Plastic Resin

Plaster and other chemical curing materials such as concrete and plastic resin may be cast using single-use waste molds as noted above, multiple-use 'piece' molds, or molds made of small rigid pieces or of flexible material such as latex rubber (which is in turn supported by an exterior mold). When casting plaster or concrete, the material surface is flat and lacks transparency. Often topical treatments are applied to the surface. For example, painting and etching can be used in a way that give the appearance of metal or stone. Alternatively, the material is altered in its initial casting process and may contain colored sand so as to give an appearance of stone. By casting concrete, rather than plaster, it is possible to create sculptures, fountains, or seating for outdoor use. A simulation of high-quality marble may be made using certain chemically-set plastic resins (for example epoxy or polyester which are thermosetting polymers) with powdered stone added for coloration, often with multiple colors worked in. The latter is a common means of making washstands, washstand tops and shower stalls, with the skilled working of multiple colors resulting in simulated staining patterns as is often found in natural marble or travertine.

Fettling

Raw castings often contain irregularities caused by seams and imperfections in the molds, as well as access ports for pouring material into the molds. The process of cutting, grinding, shaving or sanding away these unwanted bits is called "fettling". In modern times robotic processes have been developed to perform some of the more repetitive parts of the fettling process, but historically fettlers carried out this arduous work manually, and often in conditions dangerous to their health.

Fettling can add significantly to the cost of the resulting product, and designers of molds seek to minimize it through the shape of the mold, the material being cast, and sometimes by including decorative elements.

Casting Process Simulation

Casting process simulation uses numerical methods to calculate cast component quality considering mold filling, solidification and cooling, and provides a quantitative prediction of casting mechanical properties, thermal stresses and distortion. Simulation accurately describes a cast component's quality up-front before production starts. The casting rigging can be designed with respect to the required component properties. This has benefits beyond a reduction in pre-production sampling, as the precise layout of the complete casting system also leads to energy, material, and tooling savings.

The software supports the user in component design, the determination of melting practice and casting methoding through to pattern and mold making, heat treatment, and finishing. This saves costs along the entire casting manufacturing route.

Casting process simulation was initially developed at universities starting from the early '70s, mainly in Europe and in the U.S., and is regarded as the most important innovation in casting technology over the last 50 years. Since the late '80s, commercial programs (such as AutoCAST and MAGMA) are available which make it possible for foundries to gain new insight into what is happening inside the mold or die during the casting process.

SAND CASTING

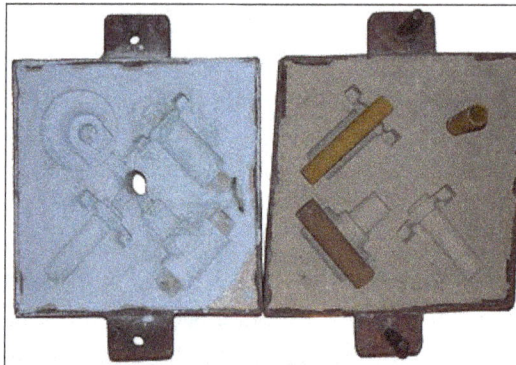

Cope & drag (top and bottom halves of a sand mold), with cores in place on the drag.

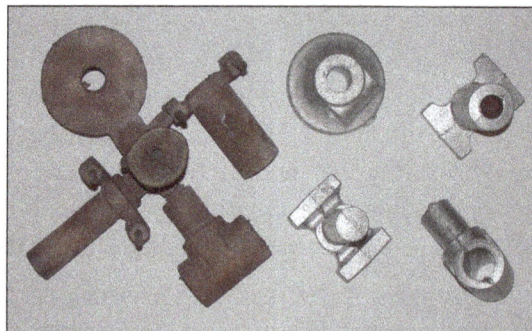

Two sets of castings (bronze and aluminium) from the above sand mold.

Sand casting, also known as sand molded casting, is a metal casting process characterized by using sand as the mold material. The term "sand casting" can also refer to an object produced via the sand casting process. Sand castings are produced in specialized factories called foundries. Over 60% of all metal castings are produced via sand casting process.

Molds made of sand are relatively cheap, and sufficiently refractory even for steel foundry use. In addition to the sand, a suitable bonding agent (usually clay) is mixed or occurs with the sand. The mixture is moistened, typically with water, but sometimes with other substances, to develop the strength and plasticity of the clay and to make the aggregate suitable for molding. The sand is typically contained in a system of frames or mold boxes known as a flask. The mold cavities and gate system are created by compacting the sand around models called patterns, by carving directly into the sand, or by 3D printing.

There are six steps in this process:

- Place a pattern in sand to create a mold.

- Incorporate the pattern and sand in a gating system.

- Remove the pattern.

- Fill the mold cavity with molten metal.

- Allow the metal to cool.

- Break away the sand mold and remove the casting.

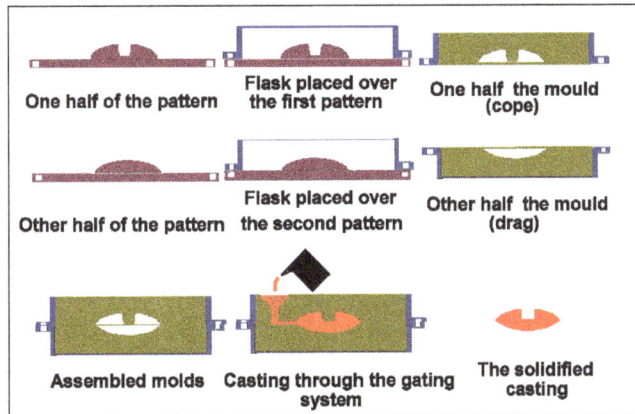

Components

Patterns

From the design, provided by a designer, a skilled pattern maker builds a pattern of the object to be produced, using wood, metal, or a plastic such as expanded polystyrene. Sand can be ground, swept or strickled into shape. The metal to be cast will contract during solidification, and this may be non-uniform due to uneven cooling. Therefore, the pattern must be slightly larger than the finished product, a difference known as contraction allowance. Different scaled rules are used for different metals, because each metal and alloy contracts by an amount distinct from all others. Patterns also have core prints that create registers within the molds into which are placed sand cores. Such cores, sometimes reinforced by wires, are used to create under-cut profiles and cavities which cannot be molded with the cope and drag, such as the interior passages of valves or cooling passages in engine blocks.

Paths for the entrance of metal into the mold cavity constitute the runner system and include the sprue, various feeders which maintain a good metal 'feed', and in-gates which attach the runner system to the casting cavity. Gas and steam generated during casting exit through the permeable sand or via risers, which are added either in the pattern itself, or as separate pieces.

Tools

In addition to patterns, the sand molder could also use tools to create the holes.

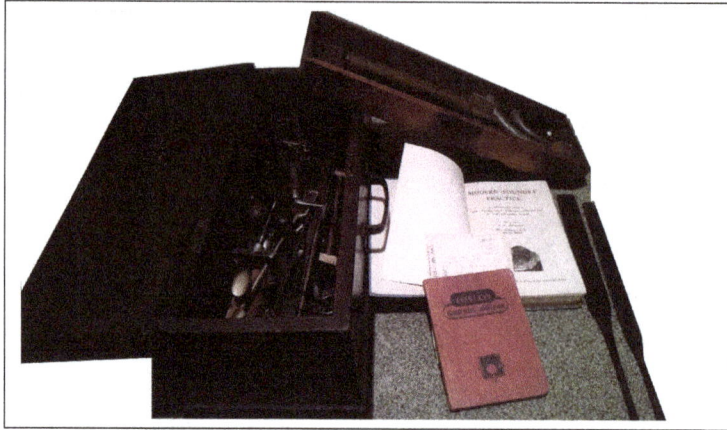

Sand Molding tools and books used in Auckland and Nelson.

Molding Box and Materials

A multi-part molding box (known as a casting flask, the top and bottom halves of which are known respectively as the cope and drag) is prepared to receive the pattern. Molding boxes are made in segments that may be latched to each other and to end closures. For a simple object—flat on one side—the lower portion of the box, closed at the bottom, will be filled with a molding sand. The sand is packed in through a vibratory process called ramming, and in this case, periodically screeded level. The surface of the sand may then be stabilized with a sizing compound. The pattern is placed on the sand and another molding box segment is added. Additional sand is rammed over and around the pattern. Finally a cover is placed on the box and it is turned and unlatched, so that the halves of the mold may be parted and the pattern with its sprue and vent patterns removed. Additional sizing may be added and any defects introduced by the removal of the pattern are corrected. The box is closed again. This forms a "green" mold which must be dried to receive the hot metal. If the mold is not sufficiently dried a steam explosion can occur that can throw molten metal about. In some cases, the sand may be oiled instead of moistened, which makes casting possible without waiting for the sand to dry. Sand may also be bonded by chemical binders, such as furane resins or amine-hardened resins.

Additive manufacturing can be used in the sand mold preparation, so that instead of the sand mold being formed via packing sand around a pattern, it is 3D-printed. This can reduce lead times for casting by obviating patternmaking. Besides replacing older methods, additive can also complement them in hybrid models, such as making a variety of AM-printed cores for a cavity derived from a traditional pattern.

Chills

To control the solidification structure of the metal, it is possible to place metal plates, chills, in the mold. The associated rapid local cooling will form a finer-grained structure and may form a somewhat harder metal at these locations. In ferrous castings, the effect is similar to quenching metals in forge work. The inner diameter of an engine cylinder is made hard by a chilling core. In other metals, chills may be used to promote directional solidification of the casting. In controlling the way a casting freezes, it is possible to prevent internal voids or porosity inside castings.

Cores

To produce cavities within the casting such as for liquid cooling in engine blocks and cylinder heads negative forms are used to produce cores. Usually sand-molded, cores are inserted into the casting box after removal of the pattern. Whenever possible, designs are made that avoid the use of cores, due to the additional set-up time, mass and thus greater cost.

With a completed mold at the appropriate moisture content, the box containing the sand mold is then positioned for filling with molten metal typically iron, steel, bronze, brass, aluminium, magnesium alloys, or various pot metal alloys, which often include lead, tin, and zinc. After being filled with liquid metal the box is set aside until the metal is sufficiently cool to be strong. The sand is then removed, revealing a rough casting that, in the case of iron or steel, may still be glowing red. In the case of metals that are significantly heavier than the casting sand, such as iron or lead, the casting flask is often covered with a heavy plate to prevent a problem known as floating the mold. Floating the mold occurs when the pressure of the metal pushes the sand above the mold cavity out of shape, causing the casting to fail.

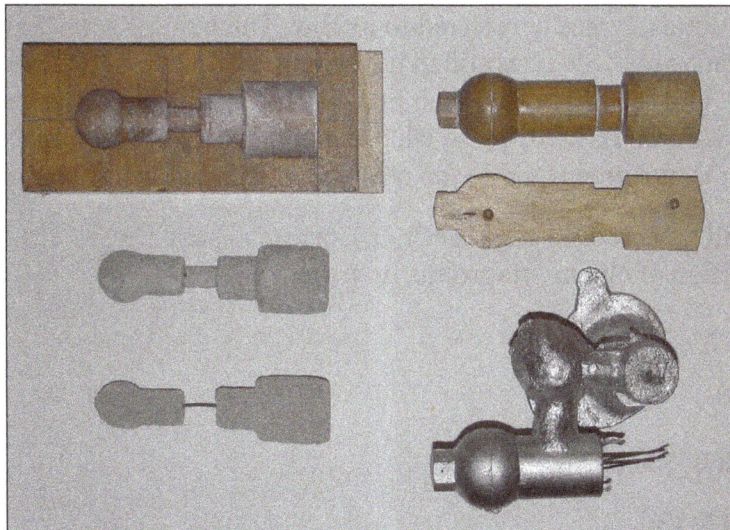

Left: Corebox, with resulting (wire reinforced) cores directly below. Right: Pattern (used with the core) and the resulting casting below (the wires are from the remains of the core).

After casting, the cores are broken up by rods or shot and removed from the casting. The metal from the sprue and risers is cut from the rough casting. Various heat treatments may be applied to relieve stresses from the initial cooling and to add hardness in the case of steel or iron, by quenching in water or oil. The casting may be further strengthened by surface compression treatment like shot peening that adds resistance to tensile cracking and smooths the rough surface. And when high precision is required, various machining operations (such as milling or boring) are made to finish critical areas of the casting. Examples of this would include the boring of cylinders and milling of the deck on a cast engine block.

Design Requirements

The part to be made and its pattern must be designed to accommodate each stage of the process, as it must be possible to remove the pattern without disturbing the molding sand and to have proper locations to receive and position the cores. A slight taper, known as draft, must be used on surfaces perpendicular to the parting line, in order to be able to remove the pattern from the mold. This requirement also applies to cores, as they must be removed from the core box in which they are formed. The sprue and risers must be arranged to allow a proper flow of metal and gasses within the mold in order to avoid an incomplete casting. Should a piece of core or mold become dislodged it may be embedded in the final casting, forming a sand pit, which may render the casting unusable. Gas pockets can cause internal voids. These may be immediately visible or may only be revealed after extensive machining has been performed. For critical applications, or where the cost of wasted effort is a factor, non-destructive testing methods may be applied before further work is performed.

Processes

In general, we can distinguish between two methods of sand casting; the first one using green sand and the second being the air set method.

Green Sand

These castings are made using sand molds formed from "wet" sand which contains water and organic bonding compounds, typically referred to as clay. The name "Green Sand" comes from the fact that the sand mold is not "set", it is still in the "green" or uncured state even when the metal is poured in the mould. Green sand is not green in color, but "green" in the sense that it is used in a wet state (akin to green wood). Contrary to what the name suggests, "green sand" is not a type of sand on its own (that is, not greensand in the geologic sense), but is rather a mixture of:

- Silica sand (SiO_2), chromite sand ($FeCr_2O_4$), or zircon sand ($ZrSiO_4$), 75 to 85%, sometimes with a proportion of olivine, staurolite, or graphite.

- bentonite (clay), 5 to 11%,

- Water, 2 to 4%,

- Inert sludge 3 to 5%,

- Anthracite (0 to 1%).

There are many recipes for the proportion of clay, but they all strike different balances between moldability, surface finish, and ability of the hot molten metal to degas. Coal, typically referred to in foundries as sea-coal, which is present at a ratio of less than 5%, partially combusts in the presence of the molten metal, leading to offgassing of organic vapors. Green sand casting for non-ferrous metals does not use coal additives, since the CO created does not prevent oxidation. Green sand for aluminum typically uses olivine sand (a mixture of the minerals forsterite and fayalite, which is made by crushing dunite rock).

The choice of sand has a lot to do with the temperature at which the metal is poured. At the temperatures that copper and iron are poured, the clay gets inactivated by the heat, in that the montmorillonite is converted to illite, which is a non-expanding clay. Most foundries do not have the very expensive equipment to remove the burned out clay and substitute new clay, so instead, those that pour iron typically work with silica sand that is inexpensive compared to the other sands. As the clay is burned out, newly mixed sand is added and some of the old sand is discarded or recycled into other uses. Silica is the least desirable of the sands, since metamorphic grains of silica sand have a tendency to explode to form sub-micron sized particles when thermally shocked during pouring of the molds. These particles enter the air of the work area and can lead to silicosis in the workers. Iron foundries spend a considerable effort on aggressive dust collection to capture this fine silica. The sand also has the dimensional instability associated with the conversion of quartz from alpha quartz to beta quartz at 680 °C (1250 °F). Often, combustible additives such as wood flour are added to create spaces for the grains to expand without deforming the mold. Olivine, chromite, etc. are therefore used because they do not have a phase transition that causes rapid expansion of the grains, as well as offering greater density, which cools the metal faster, producing finer grain structures in the metal. Since they are not metamorphic minerals, they do not have the polycrystals found in silica, and subsequently do not form hazardous sub-micron sized particles.

Air Set Method

The air set method uses dry sand bonded with materials other than clay, using a fast curing adhesive. The latter may also be referred to as no bake mold casting. When these are used, they are collectively called "air set" sand castings to distinguish them from "green sand" castings. Two types of molding sand are natural bonded (bank sand) and synthetic (lake sand); the latter is generally preferred due to its more consistent composition.

With both methods, the sand mixture is packed around a pattern, forming a mold cavity. If necessary, a temporary plug is placed in the sand and touching the pattern in order to later form a channel into which the casting fluid can be poured. Air-set molds are often formed with the help of a casting flask having a top and bottom part, termed the cope and drag. The sand mixture is tamped down as it is added around the pattern, and the final mold assembly is sometimes vibrated to compact the sand and fill any unwanted voids in the mold. Then the pattern is removed along with the channel plug, leaving the mold cavity. The casting liquid (typically molten metal) is then poured into the mold cavity. After the metal has solidified and cooled, the casting is separated from the sand mold. There is typically no mold release agent, and the mold is generally destroyed in the removal process.

The accuracy of the casting is limited by the type of sand and the molding process. Sand castings made from coarse green sand impart a rough texture to the surface, and this makes them easy to identify. Castings made from fine green sand can shine as cast but are limited by the depth to width ratio of pockets in the pattern. Air-set molds can produce castings with smoother surfaces than coarse green sand but this method is primarily chosen when deep narrow pockets in the pattern are necessary, due to the expense of the plastic used in the process. Air-set castings can typically be easily identified by the burnt color on the surface. The castings are typically shot blasted to remove that burnt color. Surfaces can also be later ground and polished, for example when making a large bell. After molding, the casting is covered with a residue of oxides, silicates and other compounds. This residue can be removed by various means, such as grinding, or shot blasting.

During casting, some of the components of the sand mixture are lost in the thermal casting process. Green sand can be reused after adjusting its composition to replenish the lost moisture and additives. The pattern itself can be reused indefinitely to produce new sand molds. The sand molding process has been used for many centuries to produce castings manually. Since 1950, partially automated casting processes have been developed for production lines.

Air set molding has many advantages. The process is designed to meet the growing demand of casting design engineers for the foundry industry and is best suited for larger, heavier and more complex castings. This process provides an excellent as-cast surface finish for products that require high aesthetic standards.

Cold Box

Uses organic and inorganic binders that strengthen the mold by chemically adhering to the sand. This type of mold gets its name from not being baked in an oven like other sand mold types. This type of mold is more accurate dimensionally than green-sand molds but is more expensive. Thus it is used only in applications that necessitate it.

No-bake Molds

No-bake molds are expendable sand molds, similar to typical sand molds, except they also contain a quick-setting liquid resin and catalyst. Rather than being rammed, the molding sand is poured into the flask and held until the resin solidifies, which occurs at room temperature. This type of molding also produces a better surface finish than other types of sand molds. Because no heat is involved it is called a cold-setting process. Common flask materials that are used are wood, metal, and plastic. Common metals cast into no-bake molds are brass, iron (ferrous), and aluminum alloys.

Vacuum Molding

Vacuum molding (V-process) is a variation of the sand casting process for most ferrous and non-ferrous metals, in which unbonded sand is held in the flask with a vacuum. The pattern is specially vented so that a vacuum can be pulled through it. A heat-softened thin sheet (0.003 to 0.008 in (0.076 to 0.203 mm)) of plastic film is draped over the pattern and a vacuum is drawn (200 to 400 mmHg (27 to 53 kPa)). A special vacuum forming flask is placed over the plastic pattern and is filled with a free-flowing sand. The sand is vibrated to compact the sand and a sprue and pouring cup are formed in the cope. Another sheet of plastic is placed over the top of the sand in the flask and a vacuum is drawn through the special flask; this hardens and strengthens the unbonded sand. The vacuum is then released on the pattern and the cope is removed. The drag is made in the same way (without the sprue and pouring cup). Any cores are set in place and the mold is closed. The molten metal is poured while the cope and drag are still under a vacuum, because the plastic vaporizes but the vacuum keeps the shape of the sand while the metal solidifies. When the metal has solidified, the vacuum is turned off and the sand runs out freely, releasing the casting.

The V-process is known for not requiring a draft because the plastic film has a certain degree of lubricity and it expands slightly when the vacuum is drawn in the flask. The process has high dimensional accuracy, with a tolerance of ±0.010 in for the first inch and ±0.002 in/in thereafter.

Cross-sections as small as 0.090 in (2.3 mm) are possible. The surface finish is very good, usually between 150 and 125 rms. Other advantages include no moisture related defects, no cost for binders, excellent sand permeability, and no toxic fumes from burning the binders. Finally, the pattern does not wear out because the sand does not touch it. The main disadvantage is that the process is slower than traditional sand casting so it is only suitable for low to medium production volumes; approximately 10 to 15,000 pieces a year. However, this makes it perfect for prototype work, because the pattern can be easily modified as it is made from plastic.

Fast Mold Making Processes

With the fast development of the car and machine building industry the casting consuming areas called for steady higher productivity. The basic process stages of the mechanical molding and casting process are similar to those described under the manual sand casting process. The technical and mental development however was so rapid and profound that the character of the sand casting process changed radically.

Mechanized Sand Molding

The first mechanized molding lines consisted of sand slingers and/or jolt-squeeze devices that compacted the sand in the flasks. Subsequent mold handling was mechanical using cranes, hoists and straps. After core setting the copes and drags were coupled using guide pins and clamped for closer accuracy. The molds were manually pushed off on a roller conveyor for casting and cooling.

Automatic High Pressure Sand Molding Lines

Increasing quality requirements made it necessary to increase the mold stability by applying steadily higher squeeze pressure and modern compaction methods for the sand in the flasks. In early fifties the high pressure molding was developed and applied in mechanical and later automatic flask lines. The first lines were using jolting and vibrations to pre-compact the sand in the flasks and compressed air powered pistons to compact the molds.

Horizontal Sand Flask Molding

In the first automatic horizontal flask lines the sand was shot or slung down on the pattern in a flask and squeezed with hydraulic pressure of up to 140 bars. The subsequent mold handling including turn-over, assembling, pushing-out on a conveyor were accomplished either manually or automatically. In the late fifties hydraulically powered pistons or multi-piston systems were used for the sand compaction in the flasks. This method produced much more stable and accurate molds than it was possible manually or pneumatically. In the late sixties mold compaction by fast air pressure or gas pressure drop over the pre-compacted sand mold was developed (sand-impulse and gas-impact). The general working principle for most of the horizontal flask line systems is shown on the sketch below.

Today there are many manufacturers of the automatic horizontal flask molding lines. The major disadvantages of these systems is high spare parts consumption due to multitude of movable parts, need of storing, transporting and maintaining the flasks and productivity limited to approximately 90–120 molds per hour.

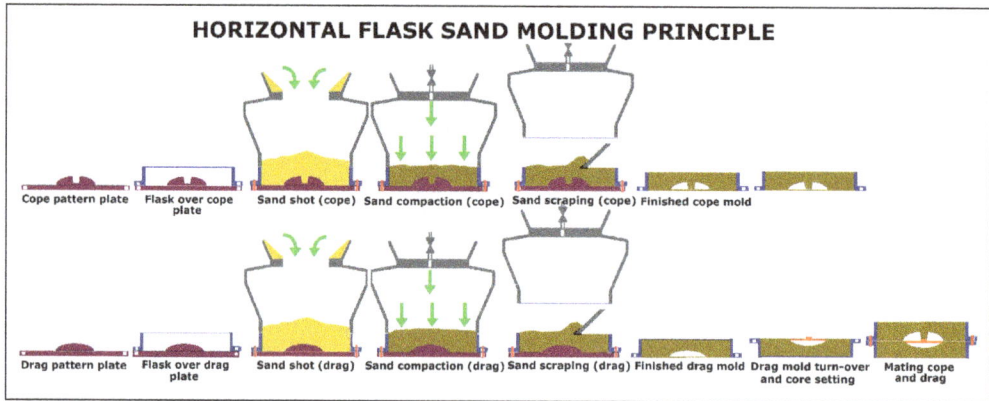

HORIZONTAL FLASK SAND MOLDING PRINCIPLE

Vertical Sand Flaskless Molding

In 1962, Dansk Industri Syndikat A/S (DISA-DISAMATIC) invented a flask-less molding process by using vertically parted and poured molds. The first line could produce up to 240 complete sand molds per hour. Today molding lines can achieve a molding rate of 550 sand molds per hour and requires only one monitoring operator. Maximum mismatch of two mold halves is 0.1 mm (0.0039 in). Although very fast, vertically parted molds are not typically used by jobbing foundries due to the specialized tooling needed to run on these machines. Cores need to be set with a core mask as opposed to by hand and must hang in the mold as opposed to being set on parting surface.

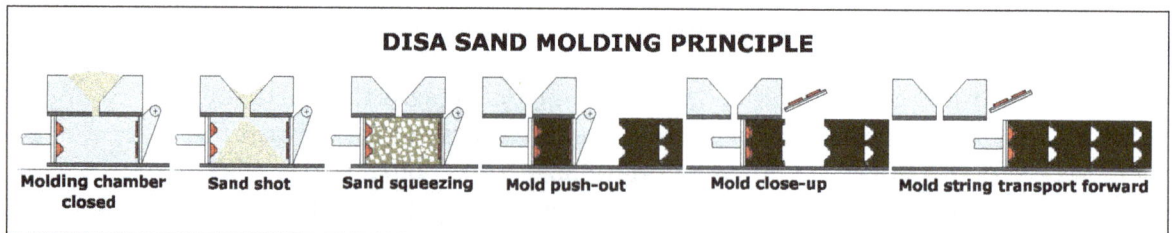

DISA SAND MOLDING PRINCIPLE

Matchplate Sand Molding

The principle of the matchplate, meaning pattern plates with two patterns on each side of the same plate, was developed and patented in 1910, fostering the perspectives for future sand molding improvements. However, first in the early sixties the American company Hunter Automated Machinery Corporation launched its first automatic flaskless, horizontal molding line applying the matchplate technology.

The method alike to the DISA's (DISAMATIC) vertical molding is flaskless, however horizontal. The matchplate molding technology is today used widely. Its great advantage is inexpensive pattern tooling, easiness of changing the molding tooling, thus suitability for manufacturing castings in short series so typical for the jobbing foundries. Modern matchplate molding machine is capable of high molding quality, less casting shift due to machine-mold mismatch (in some cases less than 0.15 mm (0.0059 in)), consistently stable molds for less grinding and improved parting line definition. In addition, the machines are enclosed for a cleaner, quieter working environment with reduced operator exposure to safety risks or service-related problems.

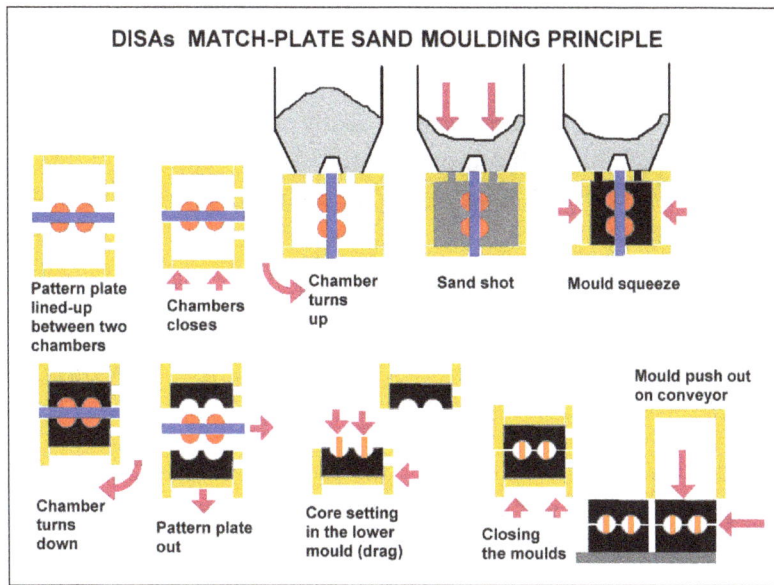

DISAs MATCH-PLATE SAND MOULDING PRINCIPLE

Mold Materials

There are four main components for making a sand casting mold: base sand, a binder, additives, and a parting compound.

Molding Sands

Molding sands, also known as foundry sands, are defined by eight characteristics: refractoriness, chemical inertness, permeability, surface finish, cohesiveness, flowability, collapsibility, and availability/cost.

Refractoriness — This refers to the sand's ability to withstand the temperature of the liquid metal being cast without breaking down. For example, some sands only need to withstand 650 °C (1,202 °F) if casting aluminum alloys, whereas steel needs a sand that will withstand 1,500 °C (2,730 °F). Sand with too low refractoriness will melt and fuse to the casting.

Chemical inertness — The sand must not react with the metal being cast. This is especially important with highly reactive metals, such as magnesium and titanium.

Permeability — This refers to the sand's ability to exhaust gases. This is important because during the pouring process many gases are produced, such as hydrogen, nitrogen, carbon dioxide, and steam, which must leave the mold otherwise casting defects, such as blow holes and gas holes, occur in the casting. Note that for each cubic centimeter (cc) of water added to the mold 16,000 cc of steam is produced.

Surface finish — The size and shape of the sand particles defines the best surface finish achievable, with finer particles producing a better finish. However, as the particles become finer (and surface finish improves) the permeability becomes worse.

Cohesiveness (or bond) — This is the ability of the sand to retain a given shape after the pattern is removed.

Flowability – The ability for the sand to flow into intricate details and tight corners without special processes or equipment.

Collapsibility — This is the ability of the sand to be easily stripped off the casting after it has solidified. Sands with poor collapsibility will adhere strongly to the casting. When casting metals that contract a lot during cooling or with long freezing temperature ranges a sand with poor collapsibility will cause cracking and hot tears in the casting. Special additives can be used to improve collapsibility.

Availability/cost — The availability and cost of the sand is very important because for every ton of metal poured, three to six tons of sand is required. Although sand can be screened and reused, the particles eventually become too fine and require periodic replacement with fresh sand.

In large castings it is economical to use two different sands, because the majority of the sand will not be in contact with the casting, so it does not need any special properties. The sand that is in contact with the casting is called facing sand, and is designed for the casting on hand. This sand will be built up around the pattern to a thickness of 30 to 100 mm (1.2 to 3.9 in). The sand that fills in around the facing sand is called backing sand. This sand is simply silica sand with only a small amount of binder and no special additives.

Types of Base Sands

Base sand is the type used to make the mold or core without any binder. Because it does not have a binder it will not bond together and is not usable in this state.

Silica Sand

Silica (SiO_2) sand is the sand found on a beach and is also the most commonly used sand. It is made by either crushing sandstone or taken from natural occurring locations, such as beaches and river beds. The fusion point of pure silica is 1,760 °C (3,200 °F), however the sands used have a lower melting point due to impurities. For high melting point casting, such as steels, a minimum of 98% pure silica sand must be used; however for lower melting point metals, such as cast iron and non-ferrous metals, a lower purity sand can be used (between 94 and 98% pure).

Silica sand is the most commonly used sand because of its great abundance, and, thus, low cost (therein being its greatest advantage). Its disadvantages are high thermal expansion, which can cause casting defects with high melting point metals, and low thermal conductivity, which can lead to unsound casting. It also cannot be used with certain basic metals because it will chemically interact with the metal, forming surface defects. Finally, it releases silica particulates during the pour, risking silicosis in foundry workers.

Olivine Sand

Olivine is a mixture of orthosilicates of iron and magnesium from the mineral dunite. Its main advantage is that it is free from silica, therefore it can be used with basic metals, such as manganese steels. Other advantages include a low thermal expansion, high thermal conductivity, and high fusion point. Finally, it is safer to use than silica, therefore it is popular in Europe.

Chromite Sand

Chromite sand is a solid solution of spinels. Its advantages are a low percentage of silica, a very high fusion point (1,850 °C (3,360 °F)), and a very high thermal conductivity. Its disadvantage is its costliness, therefore it's only used with expensive alloy steel casting and to make cores.

Zircon Sand

Zircon sand is a compound of approximately two-thirds zircon oxide (Zr_2O) and one-third silica. It has the highest fusion point of all the base sands at 2,600 °C (4,710 °F), a very low thermal expansion, and a high thermal conductivity. Because of these good properties it is commonly used when casting alloy steels and other expensive alloys. It is also used as a mold wash (a coating applied to the molding cavity) to improve surface finish. However, it is expensive and not readily available.

Chamotte Sand

Chamotte is made by calcining fire clay (Al_2O_3-SiO_2) above 1,100 °C (2,010 °F). Its fusion point is 1,750 °C (3,180 °F) and has low thermal expansion. It is the second cheapest sand, however it is still twice as expensive as silica. Its disadvantages are very coarse grains, which result in a poor surface finish, and it is limited to dry sand molding. Mold washes are used to overcome the surface finish problems. This sand is usually used when casting large steel workpieces.

Other Materials

Modern casting production methods can manufacture thin and accurate molds—of a material superficially resembling papier-mâché, such as is used in egg cartons, but that is refractory in nature—that are then supported by some means, such as dry sand surrounded by a box, during the casting process. Due to the higher accuracy it is possible to make thinner and hence lighter castings, because extra metal need not be present to allow for variations in the molds. These thin-mold casting methods have been used since the 1960s in the manufacture of cast-iron engine blocks and cylinder heads for automotive applications.

Binders

Binders are added to a base sand to bond the sand particles together (i.e. it is the glue that holds the mold together).

Clay and Water

A mixture of clay and water is the most commonly used binder. There are two types of clay commonly used: bentonite and kaolinite, with the former being the most common.

Oil

Oils, such as linseed oil, other vegetable oils and marine oils, used to be used as a binder, however due to their increasing cost, they have been mostly phased out. The oil also required careful baking at 100 to 200 °C (212 to 392 °F) to cure (if overheated, the oil becomes brittle, wasting the mold).

Resin

Resin binders are natural or synthetic high melting point gums. The two common types used are urea formaldehyde (UF) and phenol formaldehyde (PF) resins. PF resins have a higher heat resistance than UF resins and cost less. There are also cold-set resins, which use a catalyst instead of a heat to cure the binder. Resin binders are quite popular because different properties can be achieved by mixing with various additives. Other advantages include good collapsibility, low gassing, and they leave a good surface finish on the casting.

MDI (methylene diphenyl diisocyanate) is also a commonly used binder resin in the foundry core process.

Sodium Silicate

Sodium silicate [Na_2SiO_3 or $(Na_2O)(SiO_2)$] is a high strength binder used with silica molding sand. To cure the binder, carbon dioxide gas is used, which creates the following reaction:

$$Na_2O(SiO_2) + CO_2 \rightleftharpoons Na_2CO_3 + 2SiO_2 + Heat$$

The advantage to this binder is that it can be used at room temperature and is fast. The disadvantage is that its high strength leads to shakeout difficulties and possibly hot tears in the casting.

Additives

Additives are added to the molding components to improve: surface finish, dry strength, refractoriness, and "cushioning properties".

Up to 5% of reducing agents, such as coal powder, pitch, creosote, and fuel oil, may be added to the molding material to prevent wetting (prevention of liquid metal sticking to sand particles, thus leaving them on the casting surface), improve surface finish, decrease metal penetration, and burn-on defects. These additives achieve this by creating gases at the surface of the mold cavity, which prevent the liquid metal from adhering to the sand. Reducing agents are not used with steel casting, because they can carburize the metal during casting.

Up to 3% of "cushioning material", such as wood flour, saw dust, powdered husks, peat, and straw, can be added to reduce scabbing, hot tear, and hot crack casting defects when casting high temperature metals. These materials are beneficial because burn-off when the metal is poured creates tiny voids in the mold, allowing the sand particles to expand. They also increase collapsibility and reduce shakeout time.

Up to 2% of cereal binders, such as dextrin, starch, sulphite lye, and molasses, can be used to increase dry strength (the strength of the mold after curing) and improve surface finish. Cereal binders also improve collapsibility and reduce shakeout time because they burn off when the metal is poured. The disadvantage to cereal binders is that they are expensive.

Up to 2% of iron oxide powder can be used to prevent mold cracking and metal penetration, essentially improving refractoriness. Silica flour (fine silica) and zircon flour also improve refractoriness, especially in ferrous castings. The disadvantages to these additives is that they greatly reduce permeability.

Parting Compounds

To get the pattern out of the mold, prior to casting, a parting compound is applied to the pattern to ease removal. They can be a liquid or a fine powder (particle diameters between 75 and 150 micrometres (0.0030 and 0.0059 in)). Common powders include talc, graphite, and dry silica; common liquids include mineral oil and water-based silicon solutions. The latter are more commonly used with metal and large wooden patterns.

LOST-FOAM CASTING

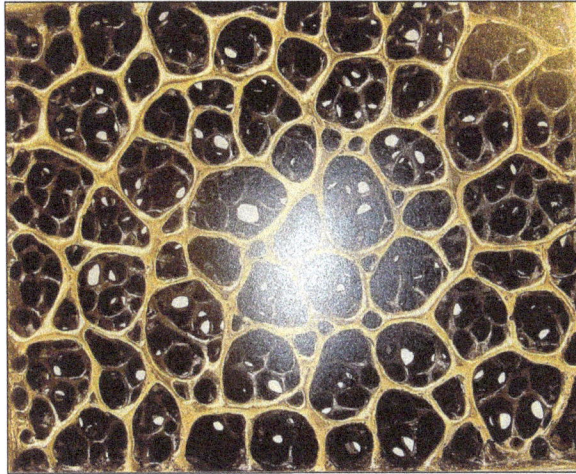
Fragment of artistic bronze casting cell.

Lost-foam casting (LFC) is a type of evaporative-pattern casting process that is similar to investment casting except foam is used for the pattern instead of wax. This process takes advantage of the low boiling point of polymer foams to simplify the investment casting process by removing the need to melt the wax out of the mold.

Process

First, a pattern is made from polystyrene foam, which can be done by many different ways. For small volume runs the pattern can be hand cut or machined from a solid block of foam; if the geometry is simple enough it can even be cut using a hot-wire foam cutter. If the volume is large, then the pattern can be mass-produced by a process similar to injection molding. Pre-expanded beads of polystyrene are injected into a preheated aluminum mold at low pressure. Steam is then applied to the polystyrene which causes it to expand more to fill the die. The final pattern is approximately 97.5% air and 2.5% polystyrene. Pre-made pouring basins, runners, and risers can be hot glued to the pattern to finish it.

Next, the foam cluster is coated with ceramic investment, also known as the refractory coating, via dipping, brushing, spraying or flow coating. After the coating dries, the cluster is placed into a flask and backed up with un-bonded sand which is compacted using a vibration table. The refractory coating captures all of the detail in the foam model and creates a barrier between the smooth foam surface and the coarse sand surface. Secondly it controls permeability, which allows the gas created by the vaporized foam pattern to escape through the coating and into the sand. Controlling permeability is a crucial step to avoid sand erosion. Finally, it forms a barrier so that molten metal does not penetrate or cause sand erosion during pouring. Once the sand is compacted, the mold is ready to be poured. Automatic pouring is commonly used in LFC, as the pouring process is significantly more critical than in conventional foundry practice.

There is no bake-out phase, as for lost-wax. The melt is poured directly into the foam-filled mold, burning out the foam as it pours. As the foam is of low density, the waste gas produced by this is relatively small and can escape through mold permeability, as for the usual outgassing control.

Details

Commonly cast metals include cast irons, aluminium alloys, steels, and nickel alloys; less frequently stainless steels and copper alloys are also cast. The size range is from 0.5 kg (1.1 lb) to several tonnes (tons). The minimum wall thickness is 2.5 mm (0.098 in) and there is no upper limit. Typical surface finishes are from 2.5 to 25 μm (100 to 1000 μin) RMS. Typical linear tolerances are ±0.005 mm/mm (0.005 in/in).

Advantages and Disadvantages

This casting process is advantageous for very complex castings that would regularly require cores. It is also dimensionally accurate, maintains an excellent surface finish, requires no draft, and has no parting lines so no flash is formed. The un-bonded sand of lost foam casting can be much simpler to maintain than green sand and resin bonded sand systems. Lost foam is generally more economical than investment casting because it involves fewer steps. Risers are not usually required due to the nature of the process; because the molten metal vaporizes the foam the first metal into the mold cools more quickly than the rest, which results in natural directional solidification. Foam is easy to manipulate, carve and glue, due to its unique properties. The flexibility of LFC often allows for consolidating the parts into one integral component; other forming processes would require the production of one or more parts to be assembled.

The two main disadvantages are that pattern costs can be high for low volume applications and the patterns are easily damaged or distorted due to their low strength. If a die is used to create the patterns there is a large initial cost.

Permanent Mold Casting

Permanent mold casting is a metal casting process that employs reusable molds ("permanent molds"), usually made from metal. The most common process uses gravity to fill the mold, however er gas pressure or a vacuum are also used. A variation on the typical gravity casting process, called slush casting, produces hollow castings. Common casting metals are aluminium, magnesium, and copper alloys. Other materials include tin, zinc, and lead alloys and iron and steel are also cast in graphite molds.

Typical products are components such as gears, splines, wheels, gear housings, pipe fittings, fuel injection housings, and automotive engine pistons.

Permanent mold casting.

Process

There are four main types of permanent mold casting: gravity, slush, low-pressure, and vacuum.

Gravity Process

The gravity process begins by preheating the mold to 150–200 °C (300–400 °F) to ease the flow and reduce thermal damage to the casting. The mold cavity is then coated with a refractory material or a mold wash, which prevents the casting from sticking to the mold and prolongs the mold life. Any sand or metal cores are then installed and the mold is clamped shut. Molten metal is then poured into the mold. Soon after solidification the mold is opened and the casting removed to reduce chances of hot tears. The process is then started all over again, but preheating is not required because the heat from the previous casting is adequate and the refractory coating should last several castings. Because this process is usually carried out on large production run work-pieces automated equipment is used to coat the mold, pour the metal, and remove the casting.

The metal is poured at the lowest practical temperature in order to minimize cracks and porosity. The pouring temperature can range greatly depending on the casting material; for instance zinc alloys are poured at approximately 370 °C (698 °F), while Gray iron is poured at approximately 1,370 °C (2,500 °F).

Mold

Molds for the casting process consist of two halves. Casting molds are usually formed from gray cast iron because it has about the best thermal fatigue resistance, but other materials include steel, bronze, and graphite. These metals are chosen because of their resistance to erosion and thermal fatigue. They are usually not very complex because the mold offers no collapsibility to compensate for shrinkage. Instead the mold is opened as soon as the casting is solidified, which prevents hot tears. Cores can be used and are usually made from sand or metal.

As stated above, the mold is heated prior to the first casting cycle and then used continuously in order to maintain as uniform a temperature as possible during the cycles. This decreases thermal fatigue, facilitates metal flow, and helps control the cooling rate of the casting metal.

Venting usually occurs through the slight crack between the two mold halves, but if this is not enough then very small vent holes are used. They are small enough to let the air escape but not the molten metal. A riser must also be included to compensate for shrinkage. This usually limits the yield to less than 60%.

Mechanical ejectors in the form of pins are used when coatings are not enough to remove casts from the molds. These pins are placed throughout the mold and usually leave small round impressions on the casting.

Slush

Slush casting is a variant of permanent molding casting to create a hollow casting or hollow cast. In the process the material is poured into the mold and allowed to cool until a shell of material forms in the mold. The remaining liquid is then poured out to leave a hollow shell. The resulting casting has good surface detail but the wall thickness can vary. The process is usually used to cast ornamental products, such as candlesticks, lamp bases, and statuary, from low-melting-point materials. A similar technique is used to make hollow chocolate figures for Easter and Christmas.

The method was developed by William Britain in 1893 for the production of lead toy soldiers. It uses less material than solid casting, and results in a lighter and less expensive product. Hollow cast figures generally have a small hole where the excess liquid was poured out.

Similarly, a process called slush molding is used in automotive dashboard manufacture, for soft-panel interiors with artificial leather, where a free-flowing (which behave like a liquid) powder plastic compound, either PVC or TPU, is poured into a hot, hollow mold and a viscous skin forms. Excess slush is then drained off, the mold is cooled, and the molded product is stripped out.

Low-pressure

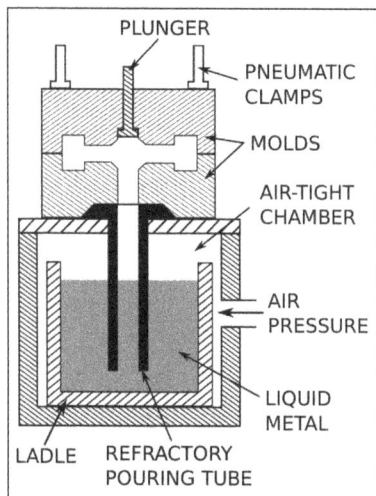

Schematic of the low-pressure permanent mold casting process.

Low-pressure permanent mold (LPPM) casting uses a gas at low pressure, usually between 3 and 15 psi (20 to 100 kPa) to push the molten metal into the mold cavity. The pressure is applied to the top of the pool of liquid, which forces the molten metal up a refractory pouring tube and finally into the bottom of the mold. The pouring tube extends to the bottom of the ladle so that the material being pushed into the mold is exceptionally clean. No risers are required because the applied pressure forces molten metal in to compensate for shrinkage. Yields are usually greater than 85% because there is no riser and any metal in the pouring tube just falls back into the ladle for reuse.

The vast majority of LPPM casting are from aluminum and magnesium, but some are copper alloys. Advantages include very little turbulence when filling the mold because of the constant pressure, which minimizes gas porosity and dross formation. Mechanical properties are about 5% better than gravity permanent mold castings. The disadvantage is that cycles times are longer than gravity permanent mold castings.

Vacuum

Vacuum permanent mold casting retains all of the advantages of LPPM casting, plus the dissolved gases in the molten metal are minimized and molten metal cleanliness is even better. The process can handle thin-walled profiles and gives an excellent surface finish. Mechanical properties are usually 10 to 15% better than gravity permanent mold castings. The process is limited in weight to 0.2 to 5 kg (0.44 to 11.02 lb).

Advantages and Disadvantages

The main advantages are the reusable mold, good surface finish, good dimensional accuracy, and high production rates. Typical tolerances are 0.4 mm for the first 25 mm (0.015 in for the first inch) and 0.02 mm for each additional centimeter (0.002 in per in); if the dimension crosses the parting line add an additional 0.25 mm (0.0098 in). Typical surface finishes are 2.5 to 7.5 µm (100–250 µin) RMS. A draft of 2 to 3° is required. Wall thicknesses are limited to 3 to 50 mm (0.12 to 1.97 in). Typical part sizes range from 100 g to 75 kg (several ounces to 150 lb). Other advantages include the ease of inducing directional solidification by changing the mold wall thickness or by heating or cooling portions of the mold. The fast cooling rates created by using a metal mold results in a finer grain structure than sand casting. Retractable metal cores can be used to create undercuts while maintaining a quick action mold.

There are three main disadvantages: high tooling cost, limited to low-melting-point metals, and short mold life. The high tooling costs make this process uneconomical for small production runs. When the process is used to cast steel or iron the mold life is extremely short. For lower melting point metals the mold life is longer but thermal fatigue and erosion usually limit the life to 10,000 to 120,000 cycles. The mold life is dependent on four factors: the mold material, the pouring temperature, the mold temperature, and the mold configuration. Molds made from gray cast iron can be more economical to produce but have short mold lives. On the other hand, molds made from H13 tool steel may have a mold life several times greater. The pouring temperature is dependent on the casting metal, but the higher the pouring temperature the shorter the mold life. A high pouring temperature can also induce shrinkage problems and create longer cycle times. If the mold temperature is too low misruns are produced, but if the mold temperature is too high then the cycle time is prolonged and mold erosion is increased. Large differences in section thickness in the mold or casting can decrease mold life as well.

MOLDING

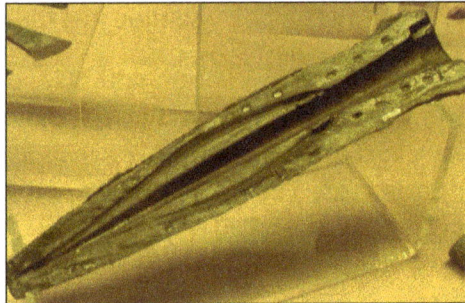

One half of a bronze mold for casting a socketed spear head dated to the period 1400-1000 BC. There are no known parallels for this mold.

Stone mold of the Bronze Age used to produce spear tips.

Ancient Greek molds, used to mass-produce clay figurines, 5th/4th century BC.
Beside them, the modern casts taken from them. On display in the Ancient
Agora Museum in Athens, housed in the Stoa of Attalus.

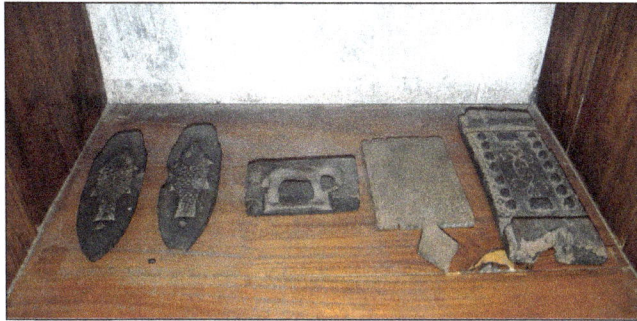

Ancient wooden molds used for jaggery & sweets, archaeological museum in Jaffna, Sri Lanka.

Molding or moulding (see spelling differences) is the process of manufacturing by shaping liquid or pliable raw material using a rigid frame called a mold or matrix. This itself may have been made using a pattern or model of the final object.

A mold or mould is a hollowed-out block that is filled with a liquid or pliable material such as plastic, glass, metal, or ceramic raw material. The liquid hardens or sets inside the mold, adopting its shape. A mold is the counterpart to a cast. The very common bi-valve molding process uses two molds, one for each half of the object. Articulated moulds have multiple pieces that come together to form the complete mold, and then disassemble to release the finished casting; they are expensive, but necessary when the casting shape has complex overhangs. Piece-molding uses a number of different molds, each creating a section of a complicated object. This is generally only used for larger and more valuable objects.

A manufacturer who makes molds is called a moldmaker. A release agent is typically used to make removal of the hardened/set substance from the mold easily. Typical uses for molded plastics include molded furniture, molded household goods, molded cases, and structural materials.

Types

There are several types of molding methods. These include:

- Blow molding.

- Powder metallurgy plus sintering.

- Compression molding.

- Extrusion molding.

- Injection molding.

- Laminating.

  o Reaction injection molding.

- Matrix molding.

- Rotational molding (or Rotomolding).

- Spin casting.

- Transfer molding.

- Thermoforming.

  o Vacuum forming, a simplified version of thermoforming.

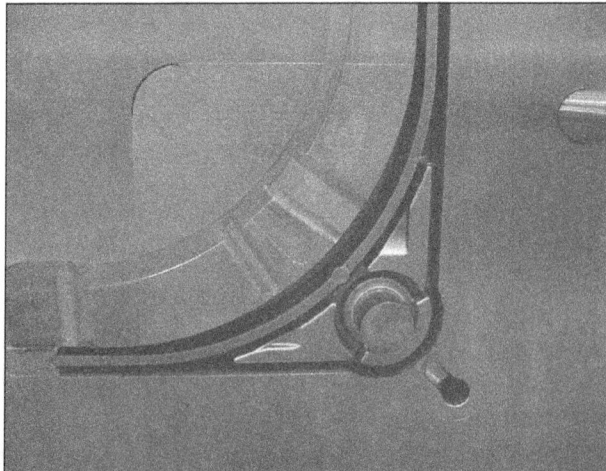

Close up of removable insert in "A" side.

"B" side of die with side pull actuators.

Insert removed from die .

BLOW MOLDING

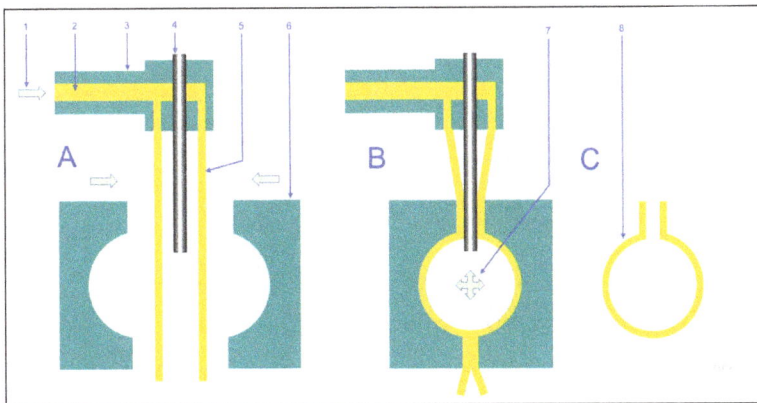

The blow molding process.

Blow molding (BrE moulding) is a specific manufacturing process by which hollow plastic parts are formed and can be joined together. It is also used for forming glass bottles or other hollow shapes.

In general, there are three main types of blow molding:

Extrusion blow molding, injection blow molding, and injection stretch blow molding.

The blow molding process begins with melting down the plastic and forming it into a parison or, in the case of injection and injection stretch blow molding (ISB), a preform. The parison is a tube-like piece of plastic with a hole in one end through which compressed air can pass.

The parison is then clamped into a mold and air is blown into it. The air pressure then pushes the plastic out to match the mold. Once the plastic has cooled and hardened the mold opens up and the part is ejected. The cost of blow molded parts is higher than that of injection-molded parts but lower than rotational molded parts.

The process principle comes from the idea of glassblowing. Enoch Ferngren and William Kopitke produced a blow molding machine and sold it to Hartford Empire Company in 1938. This was the beginning of the commercial blow molding process. During the 1940s the variety and number of products was still very limited and therefore blow molding did not take off until later. Once the variety and production rates went up the number of products created soon followed.

The technical mechanisms needed to produce hollow bodied workpieces using the blowing technique were established very early on. Because glass is very breakable, after the introduction of plastic, plastic was being used to replace glass in some cases. The first mass production of plastic bottles was done in America in 1939. Germany started using this technology a little bit later, but is currently one of the leading manufacturers of blow molding machines.

In the United States soft drink industry, the number of plastic containers went from zero in 1977 to ten billion pieces in 1999. Today, an even greater number of products are blown and it is expected to keep increasing.

For amorphous metals, also known as bulk metallic glasses, blow molding has been recently demonstrated under pressures and temperatures comparable to plastic blow molding.

Typologies

Extrusion Blow Molding

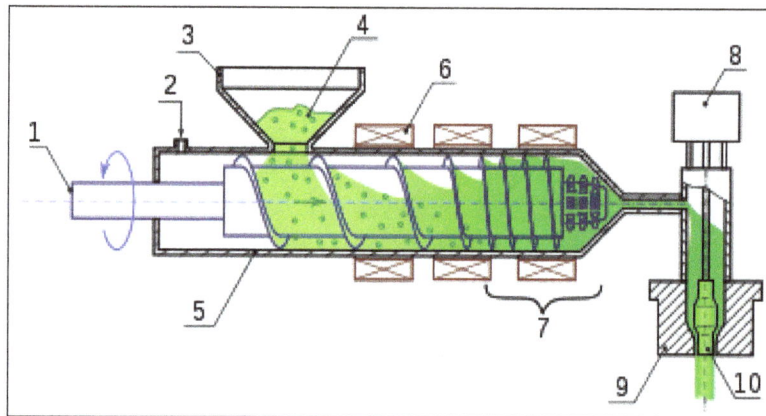

Extrusion blow molding.

In extrusion blow molding (EBM), plastic is melted and extruded into a hollow tube (a parison). This parison is then captured by closing it into a cooled metal mold. Air is then blown into the parison, inflating it into the shape of the hollow bottle, container, or part. After the plastic has cooled sufficiently, the mold is opened and the part is ejected. Continuous and Intermittent are two variations of Extrusion Blow Molding. In continuous extrusion blow molding the parison is extruded continuously and the individual parts are cut off by a suitable knife. In Intermittent blow molding there are two processes: straight intermittent is similar to injection molding whereby the screw turns, then stops and pushes the melt out. With the accumulator method, an accumulator gathers melted plastic and when the previous mold has cooled and enough plastic has accumulated, a rod pushes the melted plastic and forms the parison. In this case the screw may turn continuously or intermittently. With continuous extrusion the weight of the parison drags the parison and makes calibrating the wall thickness difficult. The accumulator head or reciprocating screw methods use hydraulic systems to push the parison out quickly reducing the effect of the weight and allowing precise control over the wall thickness by adjusting the die gap with a parison programming device.

EBM processes may be either continuous (constant extrusion of the parison) or intermittent. Types of EBM equipment may be categorized as follows:

Continuous Extrusion Equipment

- Rotary wheel blow molding systems.
- Shuttle machinery.

Intermittent Extrusion Machinery

- Reciprocating screw machinery.
- Accumulator head machinery.

Examples of parts made by the EBM process include most polyethylene hollow products, milk bottles, shampoo bottles, automotive ducting, watering cans and hollow industrial parts such as drums.

Advantages of blow molding include: Low tool and die cost; fast production rates; ability to mold complex part. Handles can be incorporated in the design.

Disadvantages of blow molding include: Limited to hollow parts, low strength, to increase barrier properties multilayer parisons of different materials are used thus not recyclable. To make wide neck jars spin trimming is necessary.

Spin Trimming

Containers such as jars often have an excess of material due to the molding process. This is trimmed off by spinning a knife around the container which cuts the material away. This excess plastic is then recycled to create new moldings. Spin Trimmers are used on a number of materials, such as PVC, HDPE and PE+LDPE. Different types of the materials have their own physical characteristics affecting trimming. For example, moldings produced from amorphous materials are much more difficult to trim than crystalline materials. Titanium coated blades are often used rather than standard steel to increase life by a factor of 30 times.

Injection Blow Molding

The process of injection blow molding (IBM) is used for the production of hollow glass and plastic objects in large quantities. In the IBM process, the polymer is injection molded onto a core pin; then the core pin is rotated to a blow molding station to be inflated and cooled. This is the least-used of the three blow molding processes, and is typically used to make small medical and single serve bottles. The process is divided into three steps: injection, blowing and ejection.

The injection blow molding machine is based on an extruder barrel and screw assembly which melts the polymer. The molten polymer is fed into a hot runner manifold where it is injected through nozzles into a heated cavity and core pin. The cavity mold forms the external shape and is clamped around a core rod which forms the internal shape of the preform. The preform consists of a fully formed bottle/jar neck with a thick tube of polymer attached, which will form the body. similar in appearance to a test tube with a threaded neck.

The preform mold opens and the core rod is rotated and clamped into the hollow, chilled blow mold. The end of the core rod opens and allows compressed air into the preform, which inflates it to the finished article shape.

After a cooling period the blow mold opens and the core rod is rotated to the ejection position. The finished article is stripped off the core rod and as an option can be leak-tested prior to packing. The preform and blow mold can have many cavities, typically three to sixteen depending on the article size and the required output. There are three sets of core rods, which allow concurrent preform injection, blow molding and ejection.

Advantages: It produces an injection molded neck for accuracy.

Disadvantages: Only suits small capacity bottles as it is difficult to control the base centre during blowing. No increase in barrier strength as the material is not biaxially stretched. Handles can't be incorporated.

Injection Stretch Blow Molding Process

This has two main different methods, namely Single-stage and two-stage process. Single-stage process is again broken down into 3-station and 4-station machines. In the two-stage injection stretch blow molding process, the plastic is first molded into a "preform" using the injection molding process. These preforms are produced with the necks of the bottles, including threads (the "finish") on one end. These preforms are packaged, and fed later (after cooling) into a reheat stretch blow molding machine. In the ISB process, the preforms are heated (typically using infrared heaters) above their glass transition temperature, then blown using high-pressure air into bottles using metal blow molds. The preform is always stretched with a core rod as part of the process.

Advantages: Very high volumes are produced. Little restriction on bottle design. Preforms can be sold as a completed item for a third party to blow. Is suitable for cylindrical, rectangular or oval bottles.

Disadvantages: High capital cost. Floor space required is high, although compact systems have become available.

In the single-stage process both preform manufacture and bottle blowing are performed in the same machine. The older 4-station method of injection, reheat, stretch blow and ejection is more costly than the 3-station machine which eliminates the reheat stage and uses latent heat in the preform, thus saving costs of energy to reheat and 25% reduction in tooling. The process explained: Imagine the molecules are small round balls, when together they have large air gaps and small surface contact, by first stretching the molecules vertically then blowing to stretch horizontally the biaxial stretching makes the molecules a cross shape. These "crosses" fit together leaving little space as more surface area is contacted thus making the material less porous and increasing barrier strength against permeation. This process also increases the strength to be ideal for filling with carbonated drinks.

Advantages

Highly suitable for low volumes and short runs. As the preform is not released during the entire process the preform wall thickness can be shaped to allow even wall thickness when blowing rectangular and non-round shapes.

Disadvantages

Restrictions on bottle design. Only a champagne base can be made for carbonated bottles.

THERMOFORMING

Thermoforming is a manufacturing process where a plastic sheet is heated to a pliable forming temperature, formed to a specific shape in a mold, and trimmed to create a usable product. The sheet, or "film" when referring to thinner gauges and certain material types, is heated in an oven to a high-enough temperature that permits it to be stretched into or onto a mold and cooled to a finished shape. Its simplified version is vacuum forming.

In its simplest form, a small tabletop or lab size machine can be used to heat small cut sections of plastic sheet and stretch it over a mold using vacuum. This method is often used for sample and prototype parts. In complex and high-volume applications, very large production machines are utilized to heat and form the plastic sheet and trim the formed parts from the sheet in a continuous high-speed process and can produce many thousands of finished parts per hour depending on the machine and mold size and the size of the parts being formed.

Thermoforming differs from injection molding, blow molding, rotational molding and other forms of processing plastics. Thin-gauge thermoforming is primarily the manufacture of disposable cups, containers, lids, trays, blisters, clamshells, and other products for the food, medical, and general retail industries. Thick-gauge thermoforming includes parts as diverse as vehicle door and dash panels, refrigerator liners, utility vehicle beds and plastic pallets.

In the most common method of high-volume, continuous thermoforming of thin-gauge products, plastic sheet is fed from a roll or from an extruder into a set of indexing chains that incorporate pins, or spikes, that pierce the sheet and transport it through an oven for heating to forming temperature. The heated sheet then indexes into a form station where a mating mold and pressure-box close on the sheet, with vacuum then applied to remove trapped air and to pull the material into or onto the mold along with pressurized air to form the plastic to the detailed shape of the mold. (Plug-assists are typically used in addition to vacuum in the case of taller, deeper-draw formed parts in order to provide the needed material distribution and thicknesses in the finished parts.) After a short form cycle, a burst of reverse air pressure is actuated from the vacuum side of the mold as the form tooling opens, commonly referred to as air-eject, to break the vacuum and assist the formed parts off of, or out of, the mold. A stripper plate may also be utilized on the mold as it opens for ejection of more detailed parts or those with negative-draft, undercut areas. The sheet containing the formed parts then indexes into a trim station on the same machine, where a die cuts the parts from the remaining sheet web or indexes into a separate trim press where the formed parts are trimmed. The sheet web remaining after the formed parts are trimmed is typically wound onto a take-up reel or fed into an inline granulator for recycling.

Most thermoforming companies recycle their scrap and waste plastic, either by compressing in a baling machine or by feeding into a granulator (grinder) and producing ground flake, for sale to reprocessing companies or re-use in their own facility. Frequently, scrap and waste plastic from the thermoforming process is converted back into extruded sheet for forming again.

Thin-gauge and Heavy-gauge (Thick) Thermoforming

There are two general thermoforming process categories. Sheet thickness less than 1.5 mm (0.060 inches) is usually delivered to the thermoforming machine from rolls or from a sheet extruder. Thin-gauge roll-fed or inline extruded thermoforming applications are dominated by rigid or semi-rigid disposable packaging. Sheet thicknesses greater than 3 mm (0.120 inches) are usually delivered to the forming machine by hand or an auto-feed method already cut to final dimensions. Heavy, or thick-gauge, cut sheet thermoforming applications are primarily used as permanent structural components. There is a small but growing medium-gauge market that forms sheet 1.5 mm to 3 mm in thickness.

Heavy-gauge forming utilizes the same basic process as continuous thin-gauge sheet forming, typically draping the heated plastic sheet over a mold. Many heavy-gauge forming applications use vacuum only in the form process, although some use two halves of mating form tooling and include air pressure to help form. Aircraft windscreens and machine gun turret windows spurred the advance of heavy-gauge forming technology during World War II. Heavy-gauge parts are used as cosmetic surfaces on permanent structures such as kiosks, automobiles, trucks, medical equipment, material handling equipment, refrigerators, spas, and shower enclosures, and electrical and electronic equipment. Unlike most thin-gauge thermoformed parts, heavy-gauge parts are often hand-worked after forming for trimming to final shape or for additional drilling, cutting, or finishing, depending on the product. Heavy-gauge products typically are of a "permanent" end use nature, while thin-gauge parts are more often designed to be disposable or recyclable and are primarily used to package or contain a food item or product. Heavy-gauge thermoforming is typically used for production quantities of 250 to 3000 annually, with lower tooling costs and faster product development than competing plastic technologies like injection molding.

Engineering

Thermoforming has benefited from applications of engineering technology, although the basic forming process is very similar to what was invented many years ago. Microprocessor and computer controls on more modern machinery allow for greatly increased process control and repeatability of same-job setups from one production run with the ability to save oven heater and process timing settings between jobs. The ability to place formed sheet into an inline trim station for more precise trim registration has been hugely improved due to the common use of electric servo motors for chain indexing versus air cylinders, gear racks, and clutches on older machines. Electric servo motors are also used on some modern and more sophisticated forming machines for actuation of the machine platens where form and trim tooling are mounted, rather than air cylinders which have traditionally been the industry standard, giving more precise control over closing and opening speeds and timing of the tooling. Quartz and radiant-panel oven heaters generally provide more precise and thorough sheet heating over older cal-rod type heaters, and better allow for zoning of ovens into areas of adjustable heat.

A new technology, ToolVu, has been developed to provide real-time feedback on thermoformer machines. This stand-alone system connects directly to the thermoformer and utilizes multiple sensors to record production-run data in real time including air pressure, temperature, tool strain gauge and other specifications. The system sends out multiple warnings and alerts whenever pre-set production parameters are compromised during a run. This reduces machine down time, lowers startup time and decreases startup scrap.

An integral part of the thermoforming process is the tooling, which is specific to each part that is to be produced. Thin-gauge thermoforming as described above is almost always performed on in-line machines and typically requires molds, plug assists, pressure boxes and all mounting plates as well as the trim tooling and stacker parts that pertain to the job. Thick or heavy-gauge thermoforming also requires tooling specific to each part, but because the part size can be very large, the molds can be cast aluminum or some other composite material as well as machined aluminum as in thin gauge. Typically, thick-gauge parts must be trimmed on CNC routers or hand trimmed using saws or hand routers. Even the most sophisticated thermoforming machine is limited to the quality of the tooling. Some large thermoforming manufacturers choose to have design and tool making facilities in house while others will rely on outside tool-making shops to build the tooling.

Types of Molds

- Plaster of paris mold,
- Wooden mold,
- Plastic mold,
- Aluminium mold.

LAMINATION

Laminate flooring.

Lamination is the technique/process of manufacturing a material in multiple layers, so that the composite material achieves improved strength, stability, sound insulation, appearance, or other properties from the use of the differing materials. A laminate is a permanently assembled object created using heat, pressure, welding, or gluing.

Materials

There are different lamination processes, depending on the type of materials to be laminated. The materials used in laminates can be the identical or different, depending on the process and the object to be laminated.

An example of the type of laminate using different materials would be the application of a layer of plastic filmthe "laminate" on either side of a sheet of glass the laminated subject. Vehicle windshields are commonly made composites created by laminating a tough plastic film between two layers of glass. This is to prevent shards of glass detaching from the windshield in case it breaks.

Plywood is a common example of a laminate using the same material in each layer combined with epoxy. Glued and laminated dimensional timber is used in the construction industry to make beams (glued laminated timber, or Glulam), in sizes larger and stronger than those that can be obtained from single pieces of wood. Another reason to laminate wooden strips into beams is quality control, as with this method each and every strip can be inspected before it becomes part of a highly stressed component.

Electrical equipment such as transformers and motors usually use a steel laminate to form the core of the coils used to produce magnetic fields. The thin lamination reduces the signal loss due to eddy currents.

Building Materials

Examples of laminate materials include melamine adhesive countertop surfacing and plywood. Decorative laminates and some modern millwork components are produced with decorative papers with a layer of overlay on top of the decorative paper, set before pressing them with thermoprocessing into high-pressure decorative laminates. A new type of HPDL is produced using real wood veneer or multilaminar veneer as top surface. High-pressure laminates consists of laminates "molded and cured at pressures not lower than 1,000 lb per sq in.(70 kg per cm²) and more commonly in the range of 1,200 to 2,000 lb per sq in. (84 to 140 kg per cm²). Meanwhile, low pressure laminate is defined as "a plastic laminate molded and cured at pressures in general of 400 pounds per square inch (approximately 27 atmospheres or 2.8×106 pascals).

Paper

A paper sign that has been laminated so it could be used outdoors.

Corrugated fiberboard boxes are examples of laminated structures, where an inner core provides rigidity and strength, and the outer layers provide a smooth surface. A starch based adhesive is usually used.

Laminating paper products, such as photographs, can prevent them from becoming creased, faded, water damaged, wrinkled, stained, smudged, abraded, or marked by grease or fingerprints. Photo identification cards and credit cards are almost always laminated with plastic film. Boxes and other containers may be laminated using heat seal layers, extrusion coatings, pressure sensitive adhesives, UV coating, etc.

Lamination is also used in sculpture using wood or resin. An example of an artist who used lamination in his work is the American Floyd Shaman.

Laminates can be used to add properties to a surface, usually printed paper, that would not have them otherwise, such as with the use of lamination paper. Sheets of vinyl impregnated with ferro-magnetic material can allow portable printed images to bond to magnets, such as for a custom bulletin board or a visual presentations. Specially surfaced plastic sheets can be laminated over a printed image to allow them to be safely written upon, such as with dry erase markers or chalk. Multiple translucent printed images may be laminated in layers to achieve certain visual effects or to hold holographic images. Printing businesses that do commercial lamination keep a variety of laminates on hand, as the process for bonding different types is generally similar when working with thin materials.

Photo Laminators

Three types of laminators are used most often in digital imaging:

- Pouch laminators.
- Heated roll laminators.
- Cold roll laminators.

Film Types

Laminate plastic film is generally categorized into these five categories:

- Standard thermal laminating films.
- Low-temperature thermal laminating films.
- Heat set (or heat-assisted) laminating films.
- Pressure-sensitive films.
- Liquid laminate.

DIP MOULDING

Dip moulding is a manufacturing process similar to candle making in its methods. Heated metal moulds are dipped in a tank of liquid material. This material may be heated or at ambient temperature. The moulds are extracted and then put through a baking process to cure the material before the part is then stripped from the mould. Once baked the material does not return to the liquid state.

As with candle making, the mould may be dipped several times to build up layers. The mould may also be dipped into different materials to create different effects. For example, handles for hand tools may be dipped first in a hard material to provide a tight fit and durability, and then into a softer material for a comfortable grip.

Materials

The materials used are known as Plastisol and are vinyl based. There is now a wide range of these materials offering variations of hardness, colours, matt or gloss finish as well as foams and grades that look and feel like suede or rubber.

Applications

- Handles for hand tools.
- Caps.
- Plugs.
- Surgical gloves.
- Covers.

Advantages

- Low tooling costs.
- Short lead times.
- Ability to vary surface texture without the need for new tooling.
- Ability to mould complex hollow shapes.

Disadvantages

- Unable to withstand high temperatures.
- Large parts are difficult to dip mould.
- Difficult to control the dimensions of outer surfaces.

ROTATIONAL MOLDING

Rotational Molding (BrE moulding) involves a heated hollow mold which is filled with a charge or shot weight of material. It is then slowly rotated (usually around two perpendicular axes), causing the softened material to disperse and stick to the walls of the mold. In order to maintain even thickness throughout the part, the mold continues to rotate at all times during the heating phase and to avoid sagging or deformation also during the cooling phase. The process was applied to plastics in the 1950s but in the early years was little used because it was a slow process restricted to a small number of plastics. Over time, improvements in process control and developments with plastic powders have resulted in a significant increase in usage.

Rotocasting (also known as rotacasting), by comparison, uses self-curing resins in an unheated mould, but shares slow rotational speeds in common with rotational molding. Spin casting should not be confused with either, utilizing self-curing resins or white metal in a high-speed centrifugal casting machine.

A three-motor powered (tri-power) rotational-molding or spin-casting machine.

Equipment and Tooling

Rotational molding machines are made in a wide range of sizes. They normally consist of molds, an oven, a cooling chamber, and mold spindles. The spindles are mounted on a rotating axis, which provides a uniform coating of the plastic inside each mold.

Molds (or tooling) are either fabricated from welded sheet steel or cast. The fabrication method is often driven by part size and complexity; most intricate parts are likely made out of cast tooling. Molds are typically manufactured from stainless steel or aluminum. Aluminum molds are usually much thicker than an equivalent steel mold, as it is a softer metal. This thickness does not affect cycle times significantly since aluminum's thermal conductivity is many times greater than steel. Due to the need to develop a model prior to casting, cast molds tend to have additional costs associated with the manufacturing of the tooling, whereas fabricated steel or aluminum molds, particularly when used for less complex parts, are less expensive. However, some molds contain both aluminum and steel. This allows for variable thicknesses in the walls of the product. While this process is not as precise as injection molding, it does provide the designer with more options. The aluminum addition to the steel provides more heat capacity, causing the melt-flow to stay in a fluid state for a longer period.

Standard Setup and Equipment for Rotational Molding

Normally all rotation molding systems have a number of parts including molds, oven, cooling chamber and mold spindles. The molds are used to create the part, and are typically made of aluminium. The quality and finish of the product is directly related to the quality of the mold being used. The oven is used to heat the part while also rotating the part to form the part desired. The cooling chamber is where the part is placed until it cools, and the spindles are mounted to rotate and provide a uniform coat of plastic inside each mold.

Rotational Molding Machines

Rock and Roll Machine

This is a specialized machine designed mainly to produce long narrow parts. Some are of the clamshell type, thus one arm, but there are also shuttle-type Rock & Roll machines, with two arms. Each arm rotates or rolls the mold 360 degrees in one direction and at the same time tips and rocks the mold 45 degrees above or below horizontal in the other direction. Newer machines use forced hot air to heat the mold. These machines are best for large parts that have a large length-to-width ratio. Because of the smaller heating chambers, there is a saving in heating costs compared to bi-axial machines.

A Rock and Roll rotational molding machine built.

Clamshell Machine

This is a single arm rotational molding machine. The arm is usually supported by other arms on both ends. The clamshell machine heats and cools the mold in the same chamber. It takes up less space than equivalent shuttle and swing arm rotational molders. It is low in cost compared to the size of products made. It is available in smaller scales for schools interested in prototyping and for high quality models. More than one mold can be attached to the single arm.

Vertical or up and over Rotational Machine

The loading and unloading area is at the front of the machine between the heating and cooling areas. These machines vary in size between small to medium compared to other rotational machines. Vertical rotational molding machines are energy efficient due to their compact heating and cooling chambers. These machines have the same (or similar) capabilities as the horizontal carousel multi-arm machines, but take up much less space.

Shuttle Machine

Most shuttle machines have two arms that move the molds back and forth between the heating chamber and cooling station. The arms are independent of each other and they turn the molds bi-axially. In some cases, the shuttle machine has only one arm. This machine moves the mold in a linear direction in and out of heating and cooling chambers. It is low in cost for the size of product produced and the footprint is kept to a minimum compared to other types of machines. It is also available in smaller scale for schools and prototyping.

Swing Arm Machine

The swing-arm machine can have up to four arms, with a bi-axial movement. Each arm is independent from each other as it is not necessary to operate all arms at the same time. Each arm is mounted on a corner of the oven and it swings in and out of the oven. On some swing-arm machines, a pair of arms is mounted on the same corner, thus a four-arm machine has two pivot points. These machines are very useful for companies that have long cooling cycles or require a lot of time to demold parts, compared to the cook time. It is a lot easier to schedule maintenance work or try to run a new mold without interrupting production on the other arms of the machine.

Carousel Machine

A Carousel machine with four independent arms.

This is one of the most common bi-axial machines in the industry. It can have up to 4 arms and six stations and it comes in a wide range of sizes. The machine comes in two different models, fixed and independent. A fixed-arm carousel consists of 3 fixed arms that must move together. One arm will be in the heating chamber while the other is in the cooling chamber and the other in the loading/reloading area. The fixed-arm carousel works well when working with identical cycle times on each arm. The independent-arm carousel machine is available with 3 or 4 arms that can move separately from the other. This allows for different size molds, with different cycle times and thickness needs.

Production Process

The rotational molding process is a high-temperature, low-pressure plastic-forming process that uses heat and biaxial rotation (i.e., angular rotation on two axes) to produce hollow, one-piece parts. Critics of the process point to its long cycle times only one or two cycles an hour can typically occur, as opposed to other processes such as injection molding, where parts can be made in a few seconds. The process does have distinct advantages. Manufacturing large, hollow parts such as oil tanks is much easier by rotational molding than any other method. Rotational molds are significantly cheaper than other types of mold. Very little material is wasted using this process, and excess material can often be re-used, making it a very economically and environmentally viable manufacturing process.

Unloading a molded polyethylene tank in a Shuttle machine.

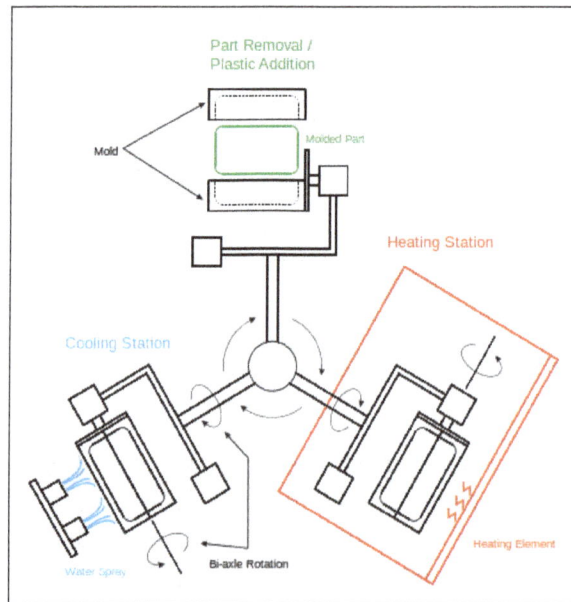
Rotational Molding Process.

The rotational molding process consists of four distinct phases:

- Loading a measured quantity of polymer (usually in powder form) into the mold.

- Heating the mold in an oven while it rotates, until all the polymer has melted and adhered to the mold wall. The hollow part should be rotated through two or more axes, rotating at different speeds, in order to avoid the accumulation of polymer powder. The length of time the mold spends in the oven is critical: too long and the polymer will degrade, reducing impact strength. If the mold spends too little time in the oven, the polymer melt may be incomplete. The polymer grains will not have time to fully melt and coalesce on the mold wall, resulting in large bubbles in the polymer. This has an adverse effect on the mechanical properties of the finished product.

- Cooling the mold, usually by fan. This stage of the cycle can be quite lengthy. The polymer must be cooled so that it solidifies and can be handled safely by the operator. This typically takes tens of minutes. The part will shrink on cooling, coming away from the mold, and facilitating easy removal of the part. The cooling rate must be kept within a certain range. Very rapid cooling (for example, water spray) would result in cooling and shrinking at an uncontrolled rate, producing a warped part.

- Removal of the part.

Recent Improvements

Until recently, the process largely relied on both trial and error and the experience of the operator to determine when the part should be removed from the oven and when it was cool enough to be removed from the mold. Technology has improved in recent years, allowing the air temperature in the mold to be monitored, removing much of the guesswork from the process.

Much of the current research is into reducing the cycle time, as well as improving part quality. The most promising area is in mold pressurization. It is well known that applying a small amount of pressure internally to the mold at the correct point in the heating phase accelerates coalescence of the polymer particles during the melting, producing a part with fewer bubbles in less time than at atmospheric pressure. This pressure delays the separation of the part from the mold wall due to shrinkage during the cooling phase, aiding cooling of the part. The main drawback to this is the danger to the operator of explosion of a pressurized part. This has prevented adoption of mold pressurization on a large scale by rotomolding manufacturers.

Mold Release Agents

A good mold release agent (MRA) will allow the material to be removed quickly and effectively. Mold releases can reduce cycle times, defects, and browning of finished product. There are a number of mold release types available; they can be categorized as follows:

- Sacrificial coatings: the coating of MRA has to be applied each time because most of the MRA comes off on the molded part when it releases from the tool. Silicones are typical MRA compounds in this category.

- Semi-permanent coatings: the coating, if applied correctly, will last for a number of releases before requiring to be re-applied or touched up. This type of coating is most prevalent in today's rotational molding industry. The active chemistry involved in these coatings is typically a polysiloxane.

- Permanent coatings: most often some form of PTFE coating, which is applied to the mold. Permanent coatings avoid the need for operator application, but may become damaged by misuse.

Materials

More than 80% of all the material used is from the polyethylene family: cross-linked poly-ethylene (PEX), low-density polyethylene (LDPE), linear low-density polyethylene (LLDPE), high-density polyethylene (HDPE), and regrind. Other compounds are PVC plastisols, nylons, and polypropylene.

Order of materials most commonly used by industry:

- Polyethylene,
- Polypropylene,
- Polyvinyl chloride,
- Nylon,
- Polycarbonate.

These materials are also occasionally used (not in order of most used):

- Aluminum,
- Acrylonitrile butadiene styrene (ABS),
- Acetal,
- Acrylic,
- Epoxy,
- Fluorocarbons,
- Ionomer,
- Polybutylene,
- Polyester,
- Polystyrene,
- Polyurethane,
- Silicone,
- Various foods (especially chocolate).

Natural Materials

Recently it has become possible to use natural materials in the molding process. Through the use of real sands and stone chip, sandstone composite can be created which is 80% natural non-pro-cessed material.

Rotational molding of plaster is used to produce hollow statuettes.

Chocolate is rotationally molded to form hollow treats.

Products

Designers can select the best material for their application, including materials that meet U.S. Food and Drug Administration (FDA) requirements. Additives for weather resistance, flame retardation, or static elimination can be incorporated. Inserts, graphics, threads, handles, minor undercuts, flat surfaces without draft angles, or fine surface detail can be part of the design. Designs can also be multi-wall, either hollow or foam filled.

Products that can be manufactured using rotational molding include storage tanks, furniture, road signs and bollards, planters, pet houses, toys, bins and refuse containers, doll parts, road cones, footballs, helmets, canoes, rowing boats, tornado shelters, kayak hulls, underground cellars for vine and vegetables storage and playground slides. The process is also used to make highly specialised products, including unapproved containers for the transportation of nuclear fissile materials, anti-piracy ship protectors, seals for inflatable oxygen masks and lightweight components for the aerospace industry.

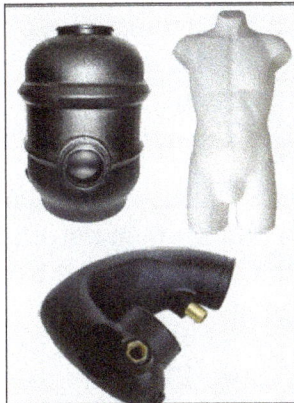

A blind brass threaded hex insert molded into a liquid storage tank.

Rotational Molded Flamingo.

Design Considerations

Designing for Rotational Molding

Another consideration is in the draft angles. These are required to remove the piece from the mold. On the outside walls, a draft angle of 1° may work (assuming no rough surface or holes). On inside walls, such as the inside of a boat hull, a draft angle of 5° may be required. This is due to shrinkage and possible part warping.

Another consideration is of structural support ribs. While solid ribs may be desirable and achievable in injection molding and other processes, a hollow rib is the best solution in rotational molding. A solid rib may be achieved through inserting a finished piece in the mold but this adds cost.

Rotational molding excels at producing hollow parts. However, care must be taken when this is done. When the depth of the recess is greater than the width there may be problems with even heating and cooling. Additionally, enough room must be left between the parallel walls to allow for the melt-flow to move properly throughout the mold. Otherwise webbing may occur. A desirable parallel wall scenario would have a gap at least three times the nominal wall thickness, with five times the nominal wall thickness being optimal. Sharp corners for parallel walls must also be considered. With angles of less than 45° bridging, webbing, and voids may occur.

Material Limitations and Considerations

Another consideration is the melt-flow of materials. Certain materials, such as nylon, will require larger radii than other materials. Additionally, the stiffness of the set material may be a factor. More structural and strengthening measures may be required when a flimsy material is used.

Wall Thickness

One benefit of rotational molding is the ability to experiment, particularly with wall thicknesses. Cost is entirely dependent on wall thickness, with thicker walls being costlier and more time consuming to produce. While the wall can have nearly any thickness, designers must remember that the thicker the wall, the more material and time will be required, increasing costs. In some cases, the plastics may significantly degrade due to extended periods at high temperature. Also, different materials have different thermal conductivity, meaning they require different times in the heating chamber and cooling chamber. Ideally, the part will be tested to use the minimum thickness required for the application. This minimum will then be established as a nominal thickness.

For the designer, while variable thicknesses are possible, a process called stop rotation is required. This process is limited in that only one side of the mold may be thicker than the others. After the mold is rotated and all the surfaces are sufficiently coated with the melt-flow, the rotation stops and the melt-flow is allowed to pool at the bottom of the mold cavity.

Wall thickness is important for corner radii as well. Large outside radii are preferable to small radii. Large inside radii are also preferable to small inside radii. This allows for a more even flow of material and a more even wall thickness. However, an outside corner is generally stronger than an inside corner.

Process: Advantages, Limitations and Material Requirements

Advantages

Rotational molding offers design advantages over other molding processes. With proper design, parts assembled from several pieces can be molded as one part, eliminating high fabrication costs. The process also has inherent design strengths, such as consistent wall thickness and strong outside corners that are virtually stress free. For additional strength, reinforcing ribs can be designed into the part. Along with being designed into the part, they can be added to the mold.

The ability to add prefinished pieces to the mold alone is a large advantage. Metal threads, internal pipes and structures, and even different colored plastics can all be added to the mold prior to the addition of plastic pellets. However, care must be taken to ensure that minimal shrinkage while cooling will not damage the part. This shrinking allows for mild undercuts and negates the need for ejection mechanisms (in most pieces).

In some cases rotational molding can be used as a feasible alternative to blow molding, this is due to the similarity in product outputs, with products such as plastic bottles and cylindrical containers, this is only effective on a smaller scale as it much more costly to blow mold regarding a small output, and with fewer resulting products rotational molding is much cheaper, due to blow molding relying on economies of scale regarding efficiency.

Another advantage lies in the molds themselves. Since they require less tooling, they can be manufactured and put into production much more quickly than other molding processes. This is especially true for complex parts, which may require large amounts of tooling for other molding processes. Rotational molding is also the process of choice for short runs and rush deliveries. The molds can be swapped quickly or different colors can be used without purging the mold. With other processes, purging may be required to swap colors.

Due to the uniform thicknesses achieved, large stretched sections are nonexistent, which makes large thin panels possible (although warping may occur). Also, there is little flow of plastic (stretching) but rather a placing of the material within the part. These thin walls also limit cost and production time.

Another cost advantage with rotational molding is the minimal amount of material wasted in production. There are no sprues or runners (as in injection molding), no off-cuts, or pinch off scrap (blow molding). What material is wasted, through scrap or failed part testing, can usually be recycled.

Limitations

Rotationally molded parts have to follow some restrictions that are different from other plastic processes. As it is a low pressure process, sometimes designers face hard to reach areas in the mold. Good quality powder may help overcome some situations, but usually the designers have to keep in mind that it is not possible to make sharp threads that would be possible with injection molding. Some products based on polyethylene can be put in the mold before filling it with the main material. This can help to avoid holes that otherwise would appear in some areas. This could also be achieved using molds with movable sections.

Another limitation lies in the molds themselves. Unlike other processes where only the product needs to be cooled before being removed, with rotational molding the entire mold must be cooled. While water cooling processes are possible, there is still a significant down time of the mold. Additionally, this increases both financial and environmental costs. Some plastics will degrade with the long heating cycles or in the process of turning them into a powder to be melted.

The stages of heating and cooling involve transfer of heat first from the hot medium to the polymer material and next from it to the cooling environment. In both cases, the process of heat transfer occurs in an unsteady regime; therefore, its kinetics attracts the greatest interest in considering these steps. In the heating stage, the heat taken from the hot gas is absorbed both by the mold and the polymer material. The rig for rotational molding usually has a relatively small wall thickness and is manufactured from metals with a high thermal conductivity (aluminum, steel). As a rule, the mold transfers much more heat than plastic can absorb; therefore, the mold temperature must vary linearly. The rotational velocity in rotational molding is rather low (4 to 20 rpm). As a result, in the first stages of the heating cycle, the charged material remains as a powder layer at the bottom of the mold. The most convenient way of changing the cycle is by applying PU sheets in hot rolled forms.

Material Requirements

Due to the nature of the process, materials selection must take into account the following:

- Due to high temperatures within the mold the plastic must have a high resistance to permanent change in properties caused by heat (high thermal stability).

- The molten plastic will come into contact with the oxygen inside the mold this can potentially lead to oxidation of the melted plastic and deterioration of the material's properties. Therefore, the chosen plastic must have a sufficient amount of antioxidant molecules to prevent such degradation in its liquid state.

- Because there is no pressure to push the plastic into the mold, the chosen plastic must be able to flow easily through the cavities of the mold. The part's design must also take into account the flow characteristics of the particular plastic chosen.

Rotational moulding is basically used for big storage tanks.

METAL INJECTION MOLDING

Metal Injection Molding (MIM) is a variation on traditional plastic injection molding that enables the fabrication of solid metal parts utilizing injection molding technology. In this process, the raw material, referred to as the feedstock, is a powder mixture of metal and polymer. For this reason, MIM is sometimes referred to as Powder Injection Molding (PIM). Using a standard injection molding machine, the powder is melted and injected into a mold, where it cools and solidifies into the shape of the desired part. Subsequent heating processes remove the unwanted polymer and produce a high-density metal part.

Metal injection molding is best suited for the high-volume production of small metal parts. As with injection molding, these parts may be geometrically complex and have thin walls and fine details. The use of metal powders enables a wide variety of ferrous and non ferrous alloys to be used and for the material properties (strength, hardness, wear resistance, corrosion resistance, etc.) to be close to those of wrought metals. Also, because the metal is not melted in the MIM process (unlike metal casting processes), high temperature alloys can be used without any negative affect on tool life. Metals commonly used for MIM parts include the following:

- Low alloy steels.
- Stainless steels.
- High-speed steels.
- Irons.
- Cobalt alloys.
- Copper alloys.
- Nickel alloys.

- Tungsten alloys.

- Titanium alloys.

Metal parts manufactured from the MIM process are found in numerous industries, including aerospace, automotive, consumer products, medical/dental, and telecommunications. MIM components can be found in cell phones, sporting goods, power tools, surgical instruments, and various electronic and optical devices.

The metal injection molding process consists of the following steps:

- Feedstock preparation - The first step is to create a powder mixture of metal and polymer. The powder metals used here are much finer (typically under 20 microns) than those used in traditional powder metallurgy processes. The powder metal is mixed with a hot thermoplastic binder, cooled, and then granulated into a homogenous feedstock in the form of pellets. The resulting feedstock is typically 60% metal and 40% polymer by volume.

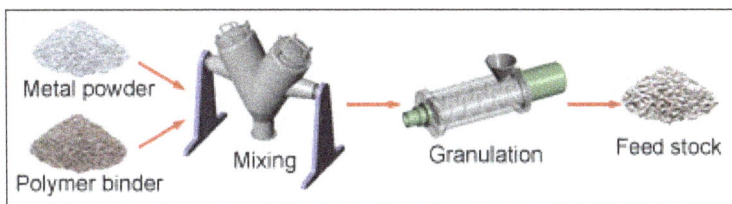
MIM Feedstock Preparation.

- Injection molding - The powder feedstock is molded using the same equipment and tooling that are used in plastic injection molding. However, the mold cavities are designed approximately 20% larger to account for the part shrinkage during sintering. In the injection molding cycle, the feedstock is melted and injected into the mold cavity, where it cools and solidifies into the shape of the part. The molded "green" part is ejected and then cleaned to remove all flash.

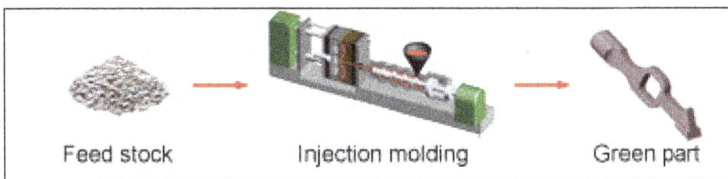
MIM Injection Molding.

- Debinding - This step removes the polymer binder from the metal. In some cases, solvent debinding is first performed in which the "green" part is placed in a water or chemical bath to dissolve most of the binder. After (on in place of) this step, thermal debinding or pre-sintering is performed. The "green" part is heated in a low temperature oven, allowing the polymer binder to be removed via evaporation. As a result, the remaining "brown" metal part will contain approximately 40% empty space by volume.

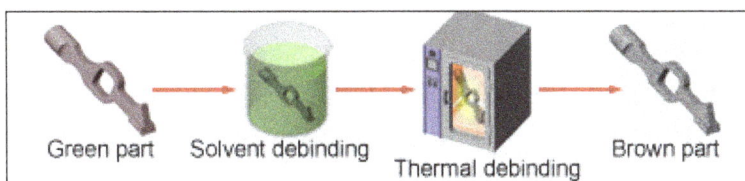
MIM Debinding.

- Sintering - The final step is to sinter the "brown" part in a high temperature furnace (up to 2500°F) in order to reduce the empty space to approximately 1-5%, resulting in a high-density (95-99%) metal part. The furnace uses an atmosphere of inert gases and attains temperatures close to 85% of the metal's melting point. This process removes pores from the material, causing the part to shrink to 75-85% of its molded size. However, this shrinkage occurs uniformly and can be accurately predicted. The resulting part retains the original molded shape with high tolerances, but is now of much greater density.

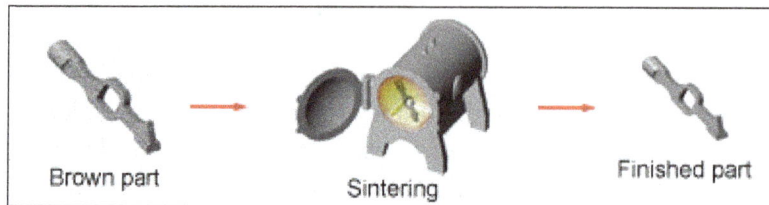

MIM Sintering.

After the sintering process, no secondary operations are required to improve tolerance or surface finish. However, just like a cast metal part, a number of secondary processes can be performed to add features, improve material properties, or assemble other components. For example, a MIM part can be machined, heat treated, or welded.

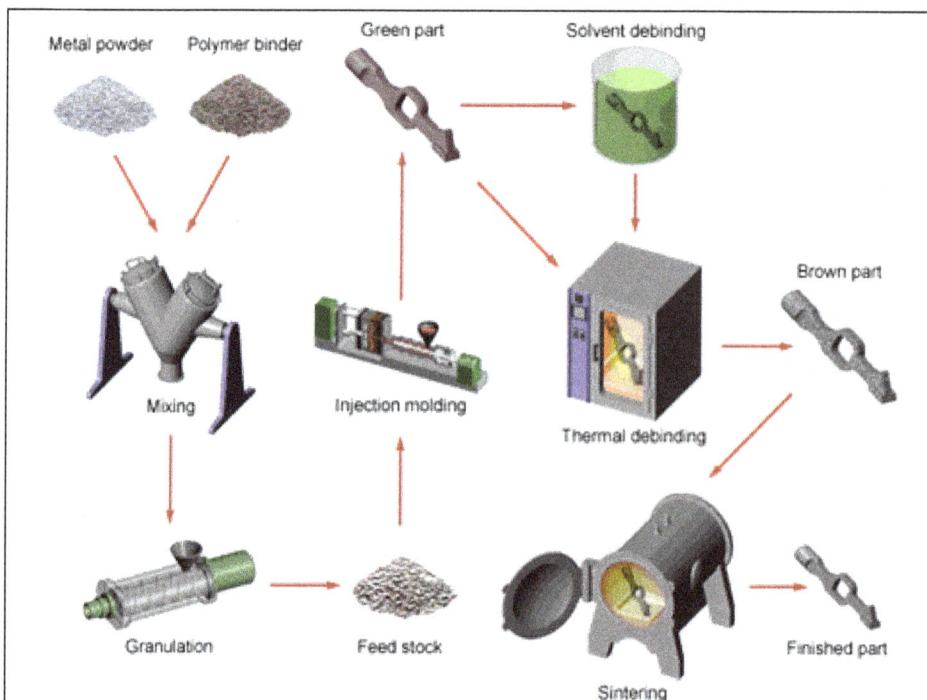

Metal Injection Molding (MIM).

Design Rules

When designing parts to be manufactured using MIM, most of the design rules for plastic injection molding still apply. However, there are some exceptions or additions, such as the following:

- Wall thickness - Just as with plastic injection molding, wall thickness should be minimized and kept uniform throughout the part. It is worth noting that in the MIM process, minimizing wall thickness not only reduces material volume and cycle time, but also reduces the debinding and sintering times as well.

- Draft - Unlike plastic injection molding, many MIM parts do not require any draft. The polymer binder used in the powder material releases more easily from the mold than most injection molded polymers. Also, MIM parts are ejected before they fully cool and shrink around the mold features because the metal powder in the mixture takes longer to cool.

- Sintering support - During sintering, MIM parts must be properly supported or they may distort as they shrink. By designing parts with flat surfaces on the same plane, standard flat support trays can be used. Otherwise, more expensive custom supports may be required.

THERMAL SPRAYING

Plasma spraying setup – a variant of thermal spraying.

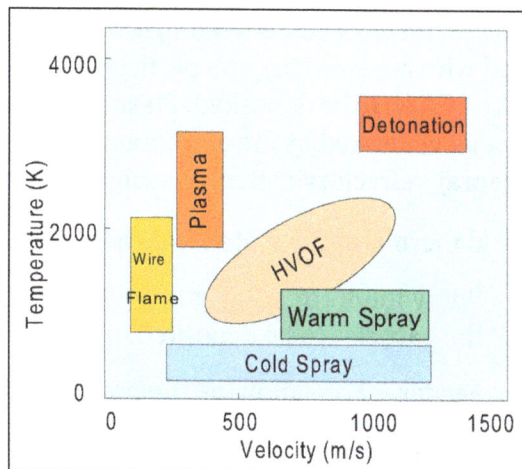

Particle temperature and velocity
for different thermal spraying processes.

Thermal spraying techniques are coating processes in which melted (or heated) materials are sprayed onto a surface. The "feedstock" (coating precursor) is heated by electrical (plasma or arc) or chemical means (combustion flame).

Thermal spraying can provide thick coatings (approx. thickness range is 20 microns to several mm, depending on the process and feedstock), over a large area at high deposition rate as compared to other coating processes such as electroplating, physical and chemical vapor deposition. Coating materials available for thermal spraying include metals, alloys, ceramics, plastics and composites. They are fed in powder or wire form, heated to a molten or semimolten state and accelerated towards substrates in the form of micrometer-size particles. Combustion or electrical arc discharge is usually used as the source of energy for thermal spraying. Resulting coatings are made by the accumulation of numerous sprayed particles. The surface may not heat up significantly, allowing the coating of flammable substances.

Coating quality is usually assessed by measuring its porosity, oxide content, macro and micro-hardness, bond strength and surface roughness. Generally, the coating quality increases with increasing particle velocities.

Several variations of thermal spraying are distinguished:

- Plasma spraying.

- Detonation spraying.

- Wire arc spraying.

- Flame spraying.

- High velocity oxy-fuel coating spraying (HVOF).

- High velocity air fuel (HVAF).

- Warm spraying.

- Cold spraying.

In classical (developed between 1910 and 1920) but still widely used processes such as flame spraying and wire arc spraying, the particle velocities are generally low (< 150 m/s), and raw materials must be molten to be deposited. Plasma spraying, developed in the 1970s, uses a high-temperature plasma jet generated by arc discharge with typical temperatures >15,000 K, which makes it possible to spray refractory materials such as oxides, molybdenum, etc.

A typical thermal spray system consists of the following:

- Spray torch (or spray gun) – The core device performing the melting and acceleration of the particles to be deposited.

- Feeder – For supplying the powder, wire or liquid to the torch through tubes.

- Media supply – Gases or liquids for the generation of the flame or plasma jet, gases for carrying the powder, etc.

- Robot – For manipulating the torch or the substrates to be coated.

- Power supply – Often standalone for the torch.

- Control console(s) – Either integrated or individual for all of the above.

Detonation Thermal Spraying Process

The detonation gun consists of a long water-cooled barrel with inlet valves for gases and powder. Oxygen and fuel (acetylene most common) are fed into the barrel along with a charge of powder. A spark is used to ignite the gas mixture, and the resulting detonation heats and accelerates the powder to supersonic velocity through the barrel. A pulse of nitrogen is used to purge the barrel after each detonation. This process is repeated many times a second. The high kinetic energy of the hot powder particles on impact with the substrate results in a buildup of a very dense and strong coating.

Plasma Spraying

Wire flame spraying.

In plasma spraying process, the material to be deposited (feedstock) typically as a powder, sometimes as a liquid, suspension or wire — is introduced into the plasma jet, emanating from a plasma torch. In the jet, where the temperature is on the order of 10,000 K, the material is melted and propelled towards a substrate. There, the molten droplets flatten, rapidly solidify and form a deposit. Commonly, the deposits remain adherent to the substrate as coatings; free-standing parts can also be produced by removing the substrate. There are a large number of technological parameters that influence the interaction of the particles with the plasma jet and the substrate and therefore the deposit properties. These parameters include feedstock type, plasma gas composition and flow rate, energy input, torch offset distance, substrate cooling, etc.

Deposit Properties

The deposits consist of a multitude of pancake-like 'splats' called lamellae, formed by flattening of the liquid droplets. As the feedstock powders typically have sizes from micrometers to above 100 micrometers, the lamellae have thickness in the micrometer range and lateral dimension from several to hundreds of micrometers. Between these lamellae, there are small voids, such as pores, cracks and regions of incomplete bonding. As a result of this unique structure, the deposits can have properties significantly different from bulk materials. These are generally mechanical properties, such as lower strength and modulus, higher strain tolerance, and lower thermal and electrical conductivity. Also, due to the rapid solidification, metastable phases can be present in the deposits.

Applications

This technique is mostly used to produce coatings on structural materials. Such coatings provide protection against high temperatures (for example thermal barrier coatings for exhaust heat management), corrosion, erosion, wear; they can also change the appearance, electrical or tribological properties of the surface, replace worn material, etc. When sprayed on substrates of various shapes and removed, free-standing parts in the form of plates, tubes, shells, etc. can be produced. It can also be used for powder processing (spheroidization, homogenization, modification of chemistry, etc.). In this case, the substrate for deposition is absent and the particles solidify during flight or in a controlled environment (e.g., water). This technique with variation may also be used to create porous structures, suitable for bone ingrowth, as a coating for medical implants. A polymer dispersion aerosol can be injected into the plasma discharge in order to create a grafting of this polymer on to a substrate surface. This application is mainly used to modify the surface chemistry of polymers.

Variations

Plasma spraying systems can be categorized by several criteria.

Plasma Jet Generation

- Direct current (DC plasma), where the energy is transferred to the plasma jet by a direct current, high-power electric arc.

- Induction plasma or RF plasma, where the energy is transferred by induction from a coil around the plasma jet, through which an alternating, radio-frequency current passes.

Plasma-forming Medium

- Gas-stabilized plasma (GSP), where the plasma forms from a gas; typically argon, hydrogen, helium or their mixtures.

- Water-stabilized plasma (WSP), where plasma forms from water (through evaporation, dissociation and ionization) or other suitable liquid.

- Hybrid plasma with combined gas and liquid stabilization, typically argon and water.

Spraying Environment

- Atmospheric plasma spraying (APS), performed in ambient air.

- Controlled atmosphere plasma spraying (CAPS), usually performed in a closed chamber, either filled with inert gas or evacuated.

- Variations of CAPS: high-pressure plasma spraying (HPPS), low-pressure plasma spraying (LPPS), the extreme case of which is vacuum plasma spraying (VPS).

- Underwater plasma spraying.

Another variation consists of having a liquid feedstock instead of a solid powder for melting, this technique is known as Solution precursor plasma spray.

Vacuum Plasma Spraying

Vacuum plasma spraying (VPS) is a technology for etching and surface modification to create porous layers with high reproducibility and for cleaning and surface engineering of plastics, rubbers and natural fibers as well as for replacing CFCs for cleaning metal components. This surface engineering can improve properties such as frictional behavior, heat resistance, surface electrical conductivity, lubricity, cohesive strength of films, or dielectric constant, or it can make materials hydrophilic or hydrophobic.

Vacuum plasma spraying.

The process typically operates at 39–120 °C to avoid thermal damage. It can induce non-thermally activated surface reactions, causing surface changes which cannot occur with molecular chemistries at atmospheric pressure. Plasma processing is done in a controlled environment inside a sealed chamber at a medium vacuum, around 13–65 Pa. The gas or mixture of gases is energized by an electrical field from DC to microwave frequencies, typically 1–500 W at 50 V. The treated components are usually electrically isolated. The volatile plasma by-products are evacuated from the chamber by the vacuum pump, and if necessary can be neutralized in an exhaust scrubber.

In contrast to molecular chemistry, plasmas employ:

- Molecular, atomic, metastable and free radical species for chemical effects.

- Positive ions and electrons for kinetic effects.

Plasma also generates electromagnetic radiation in the form of vacuum UV photons to penetrate bulk polymers to a depth of about 10 μm. This can cause chain scissions and cross-linking.

Plasmas affect materials at an atomic level. Techniques like X-ray photoelectron spectroscopy and scanning electron microscopy are used for surface analysis to identify the processes required and to judge their effects. As a simple indication of surface energy, and hence adhesion or wettability, often a water droplet contact angle test is used. The lower the contact angle, the higher the surface energy and more hydrophilic the material is.

Changing Effects with Plasma

At higher energies ionization tends to occur more than chemical dissociations. In a typical reactive gas, 1 in 100 molecules form free radicals whereas only 1 in 10^6 ionizes. The predominant effect here is the forming of free radicals. Ionic effects can predominate with selection of process parameters and if necessary the use of noble gases.

Wire Arc Spray

Wire arc spray is a form of thermal spraying where two consumable metal wires are fed independently into the spray gun. These wires are then charged and an arc is generated between them. The heat from this arc melts the incoming wire, which is then entrained in an air jet from the gun. This entrained molten feedstock is then deposited onto a substrate with the help of compressed air. This process is commonly used for metallic, heavy coatings.

Plasma Transferred Wire Arc

Plasma transferred wire Arc (PTWA) is another form of wire arc spray which deposits a coating on the internal surface of a cylinder, or on the external surface of a part of any geometry. It is predominantly known for its use in coating the cylinder bores of an engine, enabling the use of Aluminum engine blocks without the need for heavy cast iron sleeves. A single conductive wire is used as "feedstock" for the system. A supersonic plasma jet melts the wire, atomizes it and propels it onto the substrate. The plasma jet is formed by a transferred arc between a non-consumable cathode and the type of a wire. After atomization, forced air transports the stream of molten droplets onto the bore wall. The particles flatten when they impinge on the surface of the substrate, due to the high kinetic energy. The particles rapidly solidify upon contact. The stacked particles make up a high wear resistant coating. The PTWA thermal spray process utilizes a single wire as the feedstock material. All conductive wires up to and including 0.0625" (1.6mm) can be used as feedstock material, including "cored" wires. PTWA can be used to apply a coating to the wear surface of engine or transmission components to replace a bushing or bearing. For example, using PTWA to coat the bearing surface of a connecting rod offers a number of benefits including reductions in weight, cost, friction potential, and stress in the connecting rod.

High Velocity Oxygen Fuel Spraying (HVOF)

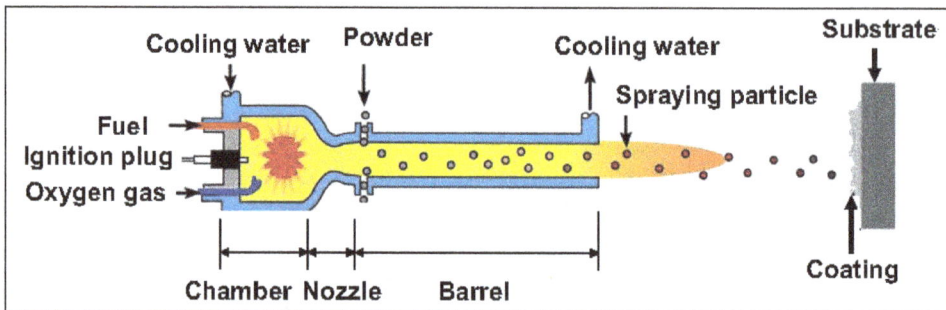

HVOF schematic.

During the 1980s, a class of thermal spray processes called high velocity oxy-fuel spraying was developed. A mixture of gaseous or liquid fuel and oxygen is fed into a combustion chamber, where they are ignited and combusted continuously. The resultant hot gas at a pressure close to 1 MPa emanates through a converging–diverging nozzle and travels through a straight section. The fuels can be gases (hydrogen, methane, propane, propylene, acetylene, natural gas, etc.) or liquids (kerosene, etc.). The jet velocity at the exit of the barrel (>1000 m/s) exceeds the speed of sound. A powder feed stock is injected into the gas stream, which accelerates the powder up to 800 m/s. The stream of hot gas and powder is directed towards the surface to be coated. The powder partially melts in the stream, and deposits upon the substrate. The resulting coating has low porosity and high bond strength.

HVOF coatings may be as thick as 12 mm (1/2"). It is typically used to deposit wear and corrosion resistant coatings on materials, such as ceramic and metallic layers. Common powders include WC-Co, chromium carbide, MCrAlY, and alumina. The process has been most successful for depositing cermet materials (WC–Co, etc.) and other corrosion-resistant alloys (stainless steels, nickel-based alloys, aluminium, hydroxyapatite for medical implants, etc.).

Cold Spraying

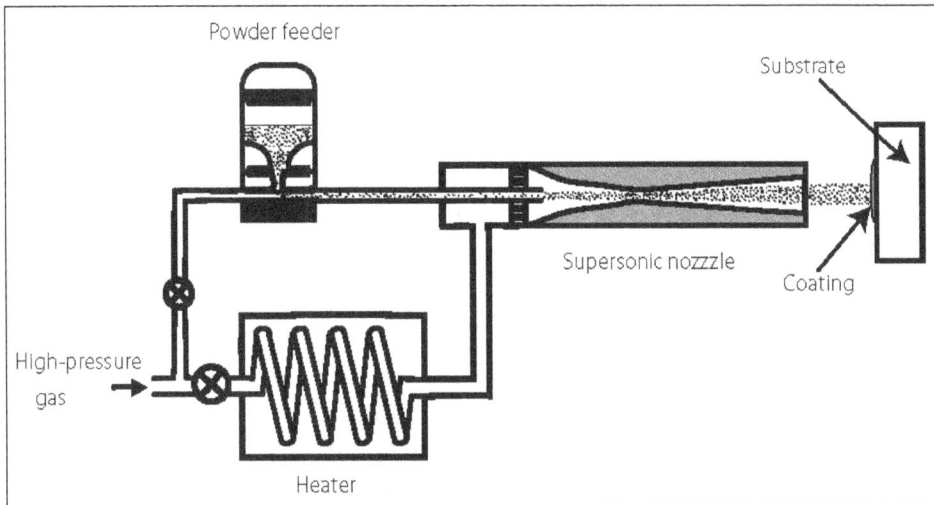

Cold spraying schematic.

Cold spraying (or gas dynamic cold spraying) was introduced to the market in the 1990s. The method was originally developed in the Soviet Union while experimenting with the erosion of the target, which was exposed to a two-phase high-velocity flow of fine powder in a wind tunnel, scientists observed accidental rapid formation of coatings.

In cold spraying, particles are accelerated to very high speeds by the carrier gas forced through a converging–diverging de Laval type nozzle. Upon impact, solid particles with sufficient kinetic energy deform plastically and bond mechanically to the substrate to form a coating. The critical velocity needed to form bonding depends on the material's properties, powder size and temperature. Metals, polymers, ceramics, composite materials and nanocrystalline powders can be deposited using cold spraying. Soft metals such as Cu and Al are best suited for cold spraying, but coating of other materials (W, Ta, Ti, MCrAlY, WC–Co, etc.) by cold spraying has been reported.

The deposition efficiency is typically low for alloy powders, and the window of process parameters and suitable powder sizes is narrow. To accelerate powders to higher velocity, finer powders (<20 micrometers) are used. It is possible to accelerate powder particles to much higher velocity using a processing gas having high speed of sound (helium instead of nitrogen). However, helium is costly and its flow rate, and thus consumption, is higher. To improve acceleration capability, nitrogen gas is heated up to about 900 °C. As a result, deposition efficiency and tensile strength of deposits increase.

Warm Spraying

Warm spraying is a novel modification of high velocity oxy-fuel spraying, in which the temperature of combustion gas is lowered by mixing nitrogen with the combustion gas, thus bringing the process closer to the cold spraying. The resulting gas contains much water vapor, unreacted hydrocarbons and oxygen, and thus is dirtier than the cold spraying. However, the coating efficiency is higher. On the other hand, lower temperatures of warm spraying reduce melting and chemical reactions of the feed powder, as compared to HVOF. These advantages are especially important for such coating materials as Ti, plastics, and metallic glasses, which rapidly oxidize or deteriorate at high temperatures.

Applications

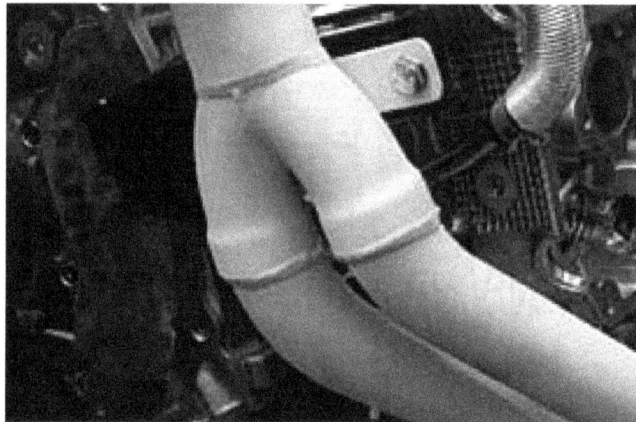

Plasma sprayed ceramic coating applied onto a part of an automotive exhaust system.

- Crankshaft reconditioning or conditioning.

- Corrosion protection.

- Fouling protection.

- Altering thermal conductivity or electrical conductivity.

- Wear control: either hardfacing (wear-resistant) or abradable coating.

- Repairing damaged surfaces.

- Temperature/oxidation protection (thermal barrier coatings).

- Medical implants.

- Production of functionally graded materials (for any of the above applications).

Limitations

Thermal spraying is a line of sight process and the bond mechanism is primarily mechanical. Thermal spray application is not compatible with the substrate if the area to which it is applied is complex or blocked by other bodies.

Safety

Thermal spraying need not be a dangerous process, if the equipment is treated with care, and correct spraying practices are followed. As with any industrial process, there are a number of hazards, of which the operator should be aware, and against which specific precautions should be taken. Ideally, equipment should be operated automatically, in enclosures specially designed to extract fumes, reduce noise levels, and prevent direct viewing of the spraying head. Such techniques will also produce coatings that are more consistent. There are occasions when the type of components being treated, or their low production levels, requires manual equipment operation. Under these conditions, a number of hazards, peculiar to thermal spraying, are experienced, in addition to those commonly encountered in production or processing industries.

Noise

Metal spraying equipment uses compressed gases, which create noise. Sound levels vary with the type of spraying equipment, the material being sprayed, and the operating parameters. Typical sound pressure levels are measured at 1 meter behind the arc.

UV Light

Combustion spraying equipment produces an intense flame, which may have a peak temperature more than 3,100 °C, and is very bright. Electric arc spraying produces ultra-violet light, which may damage delicate body tissues. Plasma also generates quite a lot of UV radiation, easily burning exposed skin and can also cause "flash burn" to the eyes. Spray booths, and enclosures, should be fitted with ultra-violet absorbent dark glass. Where this is not possible, operators, and others in the vicinity should wear protective goggles containing BS grade 6 green glass. Opaque screens should be placed around spraying areas. The nozzle of an arc pistol should never be viewed directly, unless it is certain that no power is available to the equipment.

Dust and Fumes

The atomization of molten materials produces a large amount of dust and fumes made up of very fine particles (ca. 80–95% of the particles by number <100 nm). Proper extraction facilities are vital, not only for personal safety, but to minimize entrapment of re-frozen particles in the sprayed coatings. The use of respirators, fitted with suitable filters, is strongly recommended, where equipment cannot be isolated. Certain materials offer specific known hazards:

- Finely divided metal particles are potentially pyrophoric and harmful when accumulated in the body.

- Certain materials e.g. aluminum, zinc and other base metals may react with water to evolve hydrogen. This is potentially explosive and special precautions are necessary in fume extraction equipment.

- Fumes of certain materials, notably zinc and copper alloys, have a disagreeable odour and may cause a fever-type reaction in certain individuals (known as metal fume fever). This may occur some time after spraying and usually subsides rapidly. If it does not, medical advice must be sought.

- Fumes of reactive compounds can dissociate and create harmful gasses. Respirators should be worn in these areas and gas meters should be used to monitor the air before respirators are removed.

Heat

Combustion spraying guns use oxygen and fuel gases. The fuel gases are potentially explosive. In particular, acetylene may only be used under approved conditions. Oxygen, while not explosive, will sustain combustion, and many materials will spontaneously ignite, if excessive oxygen levels are present. Care must be taken to avoid leakage, and to isolate oxygen and fuel gas supplies, when not in use.

Shock Hazards

Electric arc guns operate at low voltages (below 45 V dc), but at relatively high currents. They may be safely hand-held. The power supply units are connected to 440 V AC sources, and must be treated with caution.

DESIGN FOR MANUFACTURABILITY

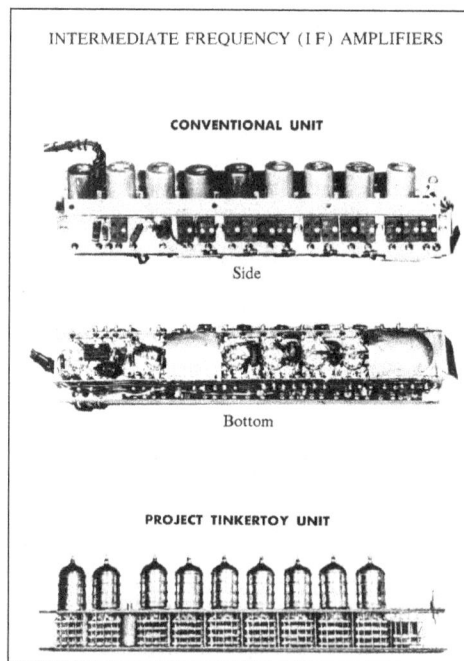

INTERMEDIATE FREQUENCY (I F) AMPLIFIERS

CONVENTIONAL UNIT

Side

Bottom

PROJECT TINKERTOY UNIT

Redesigned for manufacturability.

Design for manufacturability (also sometimes known as design for manufacturing or DFM) is the general engineering practice of designing products in such a way that they are easy to manufacture. The concept exists in almost all engineering disciplines, but the implementation differs widely depending on the manufacturing technology. DFM describes the process of designing or engineering a product in order to facilitate the manufacturing process in order to reduce its manufacturing costs. DFM will allow potential problems to be fixed in the design phase which is the least expensive place to address them. Other factors may affect the manufacturability such as the type of raw material, the form of the raw material, dimensional tolerances, and secondary processing such as finishing.

Depending on various types of manufacturing processes there are set guidelines for DFM practices. These DFM guidelines help to precisely define various tolerances, rules and common manufacturing checks related to DFM.

While DFM is applicable to the design process, a similar concept called DFSS (Design for Six Sigma) is also practiced in many organizations.

For Printed Circuit Boards (PCB)

In the PCB design process, DFM leads to a set of design guidelines that attempt to ensure manufacturability. By doing so, probable production problems may be addressed during the design stage.

Ideally, DFM guidelines take into account the processes and capabilities of the manufacturing industry. Therefore, DFM is constantly evolving.

As manufacturing companies evolve and automate more and more stages of the processes, these processes tend to become cheaper. DFM is usually used to reduce these costs. For example, if a process may be done automatically by machines (i.e. SMT component placement and soldering), such process is likely to be cheaper than doing so by hand.

For Integrated Circuits (IC)

Achieving high-yielding designs, in the state of the art VLSI technology has become an extremely challenging task due to the miniaturization as well as the complexity of leading-edge products. Here, the DFM methodology includes a set of techniques to modify the design of integrated circuits (IC) in order to make them more manufacturable, i.e., to improve their functional yield, parametric yield, or their reliability.

Background

Traditionally, in the pre-nanometer era, DFM consisted of a set of different methodologies trying to enforce some soft (recommended) design rules regarding the shapes and polygons of the physical layout of an integrated circuit. These DFM methodologies worked primarily at the full chip level. Additionally, worst-case simulations at different levels of abstraction were applied to minimize the impact of process variations on performance and other types of parametric yield loss. All these different types of worst-case simulations were essentially based on a base set of worst-case (or corner) SPICE device parameter files that were intended to represent the variability of transistor performance over the full range of variation in a fabrication process.

Taxonomy of Yield Loss Mechanisms

The most important yield loss models (YLMs) for VLSI ICs can be classified into several categories based on their nature.

Functional yield loss is still the dominant factor and is caused by mechanisms such as misprocessing (e.g., equipment-related problems), systematic effects such as printability or planarization problems, and purely random defects.

- High-performance products may exhibit parametric design marginalities caused by either process fluctuations or environmental factors (such as supply voltage or temperature).

- The test-related yield losses, which are caused by incorrect testing, can also play a significant role.

Techniques

After understanding the causes of yield loss, the next step is to make the design as resistant as possible. Techniques used for this include:

- Substituting higher yield cells where permitted by timing, power, and routability.

- Changing the spacing and width of the interconnect wires, where possible.

- Optimizing the amount of redundancy in internal memories.

- Substituting fault tolerant (redundant) vias in a design where possible.

All of these require a detailed understanding of yield loss mechanisms, since these changes trade off against one another. For example, introducing redundant vias will reduce the chance of via problems, but increase the chance of unwanted shorts. Whether this is good idea, therefore, depends on the details of the yield loss models and the characteristics of the particular design.

For CNC Machining

Objective

The objective is to design for lower cost. The cost is driven by time, so the design must minimize the time required to not just machine (remove the material), but also the set-up time of the CNC machine, NC programming, fixturing and many other activities that are dependent on the complexity and size of the part.

Set-up Time of Operations

Unless a 4th- &/or 5th-Axis is used, a CNC can only approach the part from a single direction. One side must be machined at a time (called an operation or Op). Then the part must be flipped from side to side to machine all of the features. The geometry of the features dictates whether the part must be flipped over or not. The more Ops (flip of the part), the more expensive the part because it incurs substantial "Set-up" and "Load/Unload" time.

Each operation (flip of the part) has set-up time, machine time, time to load/unload tools, time to load/unload parts, and time to create the NC program for each operation. If a part has only 1 operation, then parts only have to be loaded/unloaded once. If it has 5 operations, then load/unload time is significant.

The low hanging fruit is minimizing the number of operations (flip of the part) to create significant savings. For example, it may take only 2 minutes to machine the face of a small part, but it will take an hour to set the machine up to do it. Or, if there are 5 operations at 1.5 hours each, but only 30 minutes total machine time, then 7.5 hours is charged for just 30 minutes of machining.

Lastly, the volume (number of parts to machine) plays a critical role in amortizing the set-up time, programming time and other activities into the cost of the part. In the example above, the part in quantities of 10 could cost 7–10X the cost in quantities of 100.

Typically, the law of diminishing returns presents itself at volumes of 100–300 because set-up times, custom tooling and fixturing can be amortized into the noise.

Material Type

The most easily machined types of metals include aluminum, brass, and softer metals. As materials get harder, denser and stronger, such as steel, stainless steel, titanium, and exotic alloys, they become much harder to machine and take much longer, thus being less manufacturable. Most types of plastic are easy to machine, although additions of fiberglass or carbon fiber can reduce the machinability. Plastics that are particularly soft and gummy may have machinability problems of their own.

Material Form

Metals come in all forms. In the case of aluminum as an example, bar stock and plate are the two most common forms from which machined parts are made. The size and shape of the component may determine which form of material must be used. It is common for engineering drawings to specify one form over the other. Bar stock is generally close to 1/2 of the cost of plate on a per pound basis. So although the material form isn't directly related to the geometry of the component, cost can be removed at the design stage by specifying the least expensive form of the material.

Tolerances

A significant contributing factor to the cost of a machined component is the geometric tolerance to which the features must be made. The tighter the tolerance required, the more expensive the component will be to machine. When designing, specify the loosest tolerance that will serve the function of the component. Tolerances must be specified on a feature by feature basis. There are creative ways to engineer components with lower tolerances that still perform as well as ones with higher tolerances.

Design and Shape

As machining is a subtractive process, the time to remove the material is a major factor in determining the machining cost. The volume and shape of the material to be removed as well as how fast the tools can be fed will determine the machining time. When using milling cutters, the strength and stiffness of the tool which is determined in part by the length to diameter ratio of the tool will play the largest role in determining that speed. The shorter the tool is relative to its diameter the faster it can be fed through the material. A ratio of 3:1 (L:D) or under is optimum. If that ratio cannot be achieved, a solution like this depicted here can be used. For holes, the length to diameter ratio of the tools are less critical, but should still be kept under 10:1.

There are many other types of features which are more or less expensive to machine. Generally chamfers cost less to machine than radii on outer horizontal edges. 3D interpolation is used to create radii on edges that are not on the same plane which incur 10X the cost. Undercuts are more expensive to machine. Features that require smaller tools, regardless of L:D ratio, are more expensive.

Design for Inspection

The concept of Design for Inspection (DFI) should complement and work in collaboration with Design for Manufacturability (DFM) and Design for Assembly (DFA) to reduce product manufacturing cost and increase manufacturing practicality. There are instances when this method could cause calendar delays since it consumes many hours of additional work such as the case of the need to prepare for design review presentations and documents. To address this, it is proposed that instead of periodic inspections, organizations could adopt the framework of empowerment, particularly at the stage of product development, wherein the senior management empowers the project leader to evaluate manufacturing processes and outcomes against expectations on product performance, cost, quality and development time. Experts, however, cite the necessity for the DFI because it is crucial in performance and quality control, determining key factors such as product reliability, safety, and life cycles. For an aerospace components company, where inspection is mandatory, there is the requirement for the suitability of the manufacturing process for inspection. Here, a mechanism is adopted such as an inspectability index, which evaluates design proposals. Another example of DFI is the concept of cumulative count of conforming chart (CCC chart), which is applied in inspection and maintenance planning for systems where different types of inspection and maintenance are available.

Design for Additive Manufacturing

Additive manufacturing broadens the ability of a designer to optimize the design of a product or part (to save materials for example). Designs tailored for additive manufacturing are sometimes very different from designs tailored for machining or forming manufacturing operations.

In addition, due to some size constraints of additive manufacturing machines, sometimes the related bigger designs are split into smaller sections with self-assembly features or fasteners locators.

ADDITIVE MANUFACTURING

Additive Manufacturing (AM) is an appropriate name to describe the technologies that build 3D objects by adding layer-upon-layer of material, whether the material is plastic, metal, concrete or one day human tissue.

Common to AM technologies is the use of a computer, 3D modeling software (Computer Aided Design or CAD), machine equipment and layering material. Once a CAD sketch is produced, the AM equipment reads in data from the CAD file and lays downs or adds successive layers of liquid, powder, sheet material or other, in a layer-upon-layer fashion to fabricate a 3D object.

The term AM encompasses many technologies including subsets like 3D Printing, Rapid Prototyping (RP), Direct Digital Manufacturing (DDM), layered manufacturing and additive fabrication.

AM application is limitless. Early use of AM in the form of Rapid Prototyping focused on preproduction visualization models. More recently, AM is being used to fabricate end-use products in aircraft, dental restorations, medical implants, automobiles, and even fashion products.

While the adding of layer-upon-layer approach is simple, there are many applications of AM technology with degrees of sophistication to meet diverse needs including:

- A visualization tool in design.
- A means to create highly customized products for consumers and professionals alike.
- As industrial tooling.
- To produce small lots of production parts.
- One day production of human organs.

At MIT, where the technology was invented, projects abound supporting a range of forward-thinking applications from multi-structure concrete to machines that can build machines; while work at Contour Crafting supports structures for people to live and work in.

Some envision AM as a complement to foundational subtractive manufacturing (removing material like drilling out material) and to lesser degree forming (like forging). Regardless, AM may offer consumers and professionals alike, the accessibility to create, customize and/or repair product, and in the process, redefine current production technology.

Whether simple or sophisticated, AM is indeed AMazing and best described in the adding of layer-upon-layer, whether in plastic, metal, concrete or one day human tissue".

Examples of Additive Manufacturing (AM)

+ SLA

Very high end technology utilizing laser technology to cure layer-upon-layer of photopolymer resin (polymer that changes properties when exposed to light).

The build occurs in a pool of resin. A laser beam, directed into the pool of resin, traces the cross-section pattern of the model for that particular layer and cures it. During the build cycle, the platform on which the build is repositioned, lowering by a single layer thickness. The process repeats until the build or model is completed and fascinating to watch. Specialized material may be needed to add support to some model features. Models can be machined and used as patterns for injection molding, thermoforming or other casting processes.

+ FDM

Process oriented involving use of thermoplastic (polymer that changes to a liquid upon the application of heat and solidifies to a solid when cooled) materials injected through indexing nozzles onto a platform. The nozzles trace the cross-section pattern for each particular layer with the thermoplastic material hardening prior to the application of the next layer. The process repeats until the build or model is completed and fascinating to watch. Specialized material may be need to add support to some model features. Similar to SLA, the models can be machined or used as patterns. Very easy-to-use and cool.

+ MJM

Multi-Jet Modeling is similar to an inkjet printer in that a head, capable of shuttling back and forth (3 dimensions-x, y, z) incorporates hundreds of small jets to apply a layer of thermopolymer material, layer-by-layer.

+3DP

This involves building a model in a container filled with powder of either starch or plaster based material. An inkjet printer head shuttles applies a small amount of binder to form a layer. Upon application of the binder, a new layer of powder is sweeped over the prior layer with the application of more binder. The process repeats until the model is complete. As the model is supported by loose powder there is no need for support. Additionally, this is the only process that builds in colors.

+ SLS

Somewhat like SLA technology Selective Laser Sintering (SLS) utilizes a high powered laser to fuse small particles of plastic, metal, ceramic or glass. During the build cycle, the platform on which the build is repositioned, lowering by a single layer thickness. The process repeats until the build or model is completed. Unlike SLA technology, support material is not needed as the build is supported by unsintered material.

Design for Additive Manufacturing

Design for additive manufacturing (DfAM or DFAM) is design for manufacturability as applied to additive manufacturing (AM). It is a general type of design methods or tools whereby functional performance and/or other key product life-cycle considerations such as manufacturability, reliability, and cost can be optimized subjected to the capabilities of additive manufacturing technologies.

This concept emerges due to the enormous design freedom provided by AM technologies. To take full advantages of unique capabilities from AM processes, DFAM methods or tools are needed. Typical DFAM methods or tools includes topology optimization, design for multiscale structures (lattice or cellular structures), multi-material design, mass customization, part consolidation, and other design methods which can make use of AM-enabled features.

DFAM is not always separate from broader DFM, as the making of many objects can involve both additive and subtractive steps. Nonetheless, the name "DFAM" has value because it focuses attention on the way that commercializing AM in production roles is not just a matter of figuring out how to switch existing parts from subtractive to additive. Rather, it is about redesigning entire objects (assemblies, subsystems) in view of the newfound availability of advanced AM.

That is, it involves redesigning them because their entire earlier design—including even how, why, and at which places they were originally divided into discrete parts—was conceived within the constraints of a world where advanced AM did not yet exist. Thus instead of just modifying an existing part design to allow it to be made additively, full-fledged DFAM involves things like reimagining the overall object such that it has fewer parts or a new set of parts with substantially different boundaries and connections. The object thus may no longer be an assembly at all, or it may be an assembly with many fewer parts. Many examples of such deep-rooted practical impact of DFAM have been emerging in the 2010s, as AM greatly broadens its commercialization. For example, in 2017, GE Aviation revealed that it had used DFAM to create a helicopter engine with 16 parts instead of 900, with great potential impact on reducing the complexity of supply chains. It is this radical rethinking aspect that has led to themes such as that "DFAM requires 'enterprise-level disruption'." In other words, the disruptive innovation that AM can allow can logically extend throughout the enterprise and its supply chain, not just change the layout on a machine shop floor.

DFAM involves both broad themes (which apply to many AM processes) and optimizations specific to a particular AM process. For example, DFM analysis for stereolithography maximizes DFAM for that modality.

Additive manufacturing is defined as a material joining process, whereby a product can be directly fabricated from its 3D model, usually layer upon layer. Comparing to traditional manufacturing technologies such as CNC machining or casting, AM processes have several unique capabilities. It enables the fabrication of parts with a complex shape as well as complex material distribution. These unique capabilities significantly enlarge the design freedom for designers. However, they also bring a big challenge. Traditional Design for manufacturing (DFM) rules or guidelines deeply rooted in designers' mind and severely restrict designers to further improve product functional performance by taking advantages of these unique capabilities brought by AM processes. Moreover, traditional feature-based CAD tools are also difficult to deal with irregular geometry for the improvement of functional performance. To solve these issues, design methods or tools are needed to help designers to take full advantages of design freedom provide by AM processes. These design methods or tools can be categorized as Design for Additive Manufacturing

Methods

Topology Optimization

Topology optimization is a type of structural optimization technique which can optimize material layout within a given design space. Compared to other typical structural optimization techniques, such as size optimization or shape optimization, topology optimization can update both shape and topology of a part. However, the complex optimized shapes obtained from topology optimization are always a headache for traditional manufacturing processes such as CNC machining. To solve this issue, additive manufacturing processes can be applied to fabricate topology optimization result. However, it should be noticed, some manufacturing constraints such as minimal feature size also need to be considered during the topology optimization process. Since the topology optimization can help designers to get an optimal complex geometry for additive manufacturing, this technique can be considered one of DFAM methods.

Multiscale Structure Design

Due to the unique capabilities of AM processes, parts with multiscale complexities can be realized. This provides a great design freedom for designers to use cellular structures or lattice structures on micro or mesoscales for the preferred properties. For example, in the aerospace field, lattice structures fabricated by AM process can be used for weight reduction. In the bio-medical field, bio-implant made of lattice or cellular structures can enhance osseointegration.

Multi-material Design

Parts with multi-material or complex material distribution can be achieved by additive manufacturing processes. To help designers to take use of this advantage, several design and simulation methods has been proposed to support design a part with multiple materials or Functionally Graded Materials. These design methods also bring a challenge to traditional CAD system. Most of them can only deal with homogeneous materials now.

Design for Mass Customization

Since additive manufacturing can directly fabricate parts from products' digital model, it significantly reduces the cost and leading time of producing customized products. Thus, how to rapidly generate customized parts becomes a central issue for mass customization. Several design methods have been proposed to help designers or users to obtain the customized product in an easy way. These methods or tools can also be considered as the DFAM methods.

Parts Consolidation

Due to the constraints of traditional manufacturing methods, some complex components are usually separated into several parts for the ease of manufacturing as well as assembly. This situation has been changed by the using of additive manufacturing technologies. Some case studies have been done to shows some parts in the original design can be consolidated into one complex part and fabricated by additive manufacturing processes. This redesigning process can be called as parts consolidation. The research shows parts consolidation will not only reduce part count, it can also improve the product functional performance. The design methods which can guide designers to do part consolidation can also be regarded as a type of DFAM methods.

Lattice Structures

Lattice structures is type of cellular structures. Thanks to the free-form manufacturing capability of additive manufacturing technology, it is now possible to design complex forms. These structures were previously difficult to manufacture, hence not widely used. Lattice structures have high-strength low mass mechanical properties and multifunctionality. Lattice structures can be found in parts in the aerospace and bio-medical industries.

Lasers in Additive Manufacturing

In the year if light, no one can deny that lasers have a significant influence in fields as diverse as telecommunications, instrumentation, medicine, computing and entertainment. In manufacturing their applications include processes such as cutting, drilling, welding, bending, cladding, cleaning, marking and heat-treatment. They are being used more and at a growing range of scales; increasing powers enabling larger scale work and higher beam qualities and shorter pulse widths enabling smaller scale work. Financially, the global market for lasers is forecast at $9.7 – 11.7 billion in 2015, and expected to be $16.0 billion in 2020.

For a versatile tool with such an impressive existing portfolio of applications it may seem difficult to identify one area which is likely to be outstanding in the growth of use of the industrial laser over the next decade. However, one application has been singled out an invention that could remake the world and constitute a new industrial revolution. Alternatively, in more detail, a revolutionary technology likely to restructure supply chains, relocate production facilities and profoundly alter the geopolitical, economic, social, demographic, environmental, and security landscapes. That process is additive manufacturing (AM). Additive manufacturing began as a rapid prototyping technology, suitable for producing haptic models, and developed into what it is today: both a rapid tooling and a manufacturing technology, capable of producing fully functional parts in a wide range of materials, metallic, non-metallic and composite.

The operation of different AM systems and the relative advantages and disadvantages of the different technologies are described and reviewed in the literature. Although not holding a monopoly over AM, it is clear from the classification that lasers are required for three out of seven of the major categories and two out of three process categories with the capability to manufacture metallic components.

The development of AM systems into user-friendly, commercial units plus the need for safety, means the laser within them is not always obvious. Further, during AM of metals it is usually necessary to shield the build point from harmful oxidation, either by performing the whole operation in an inert chamber or using blown inert gas, requiring further removal of the user from the 'sharp end' of the manufacturing. To clarify exactly which types of lasers are used in AM, table lists some laser- based AM processes and systems and their lasers.

Table: Some major manufacturers of additive manufacturing equipment and their lasers.

Category	Manufacturer: systems	Laser
Directed energy deposition	Optomec Inc.: LENS series	400 - 1000 W fibre laser
	Trumpf GmbH: TruLaser series	TruDiode diode laser, up to 6 kW (600 nm)
	Oerlikon Metco Group: MetcoClad systems	1 - 6 kW diode laser
Powder bed fusion	3D Systems Inc.: ProX, sPro and ProX SLM systems	30 - 200 W CO_2 laser (for thermoplastics) 50 – 500 W fibre laser (for metals)
	EOS GmbH: EOSINT, EOS M and PRECIOUS M machines	30 W, 70 W or 2 x 50W CO_2 lasers (for thermoplastics) 200 W - 1 kW fibre laser (for metals)
	SLM Solutions GmbH: SLM systems	400 W - 2 x 1000 W fibre laser (for metals)
	Renishaw: AM250	200 W or 400 W fibre laser
Vat photopolymerisation	3D Systems Inc.: ProJet and ProX SLA ranges	Solid-state frequency tripled Nd: YVO$\square$ laser (354.7 nm), up to 1.5 W
	Envisiontec: Perfactory, Ultra, Xede, Xtreme, 3D-Bioplotter ranges	Not specified

Clearly, existing commercial AM systems currently utilise a wide range of lasers technologies. Powers range from around 1 W to 6 kW, and wavelengths from the ultraviolet (354.7 nm) to the infrared

(10.6 um). Requirements vary from process to process, but the need to match SLA lasers with the polymer absorption spectrum, the use of different lasers for different materials in the powder bed fusion bed category and the use of the shorter wavelength diode laser, despite poorer beam quality than the fibre laser for directed energy deposition (DMD), indicates absorption is a major factor for laser selection throughout. Use of systems with powers greater than 6 kW have also been demonstrated (e.g. and some industrial 'home-made' systems also exist, particularly in the DMD category.

Additive Manufacturing Production

One of the reasons use of AM is forecast to increase at such a rapid rate is that it expands the boundaries for a design engineer by offering a profoundly different approach to traditional subtractive methods. This can also allow a greater range of components to be made as a single part, reducing the material required and need for joining, by whatever means. However, these benefits can be negated by high costs.

There are a number of factors that contribute to the unit production price of an AM part, some of which can be related to the volume of production and others to the degree of complexity and customisation of the part. The former can be broken down to material cost, machine cost. Despite the greater material efficiency typically achieved, material costs are high because the raw material can be more expensive than for conventional processes. This is particularly the case for metallic parts, produced via DMD, SLM or SLS. For example, Atzeni calculated the material costs needed for a sintered aluminium alloy part to be ten times higher than those for the conventionally produced part. Machine costs are another large contributor to the unit cost, and typically the single largest cost for polymeric AM. Different studies have calculated their contribution to be between one-quarter and three-quarters of the costs of a part, with the three-quarter contribution coming when building using vat polymerisation. Build time, energy consumption and labour cost make up a smaller components of overall cost and can all be minimised by using the machines efficiently, which for most processes means effectively using the whole of the available build space to maximise solid:cavity volume ratio. Currently the amount of energy consumed when producing the same part by AM can still be greater than that when using conventional manufacturing processes, often due to the longer process times. Studies have found that SLS and 3D printing are less energy efficient than injection moulding at all but low production volumes. Sreenivasan and Bourell identified major consumers of energy in SLS of nylon materials to be chamber heaters (37 %), stepper motors for piston control (26%), roller drives (16 %), and the laser (16 %). Preparation and postprocessing costs are not included in many models, but Lindemann estimated them to be the third largest contributor to overall costs after the machine and materials. The indirect costs such as factory overheads associated with the production can be assigned in different ways, accounting for factors such as mixed part production but for any commercially viable organisation are much lower than direct costs.

Taking all these costs together, what is most evident is that in a production environment the economies of scale with AM are insignificant compared with traditional methods, leading to figure (generalised from). The current position of the 'breakeven' point due to the high AM machine, material, preparation and postprocessing costs mean large parts, high production volumes or rates, or high accuracy and surface finish quality typically render AM production more expensive than traditional manufacturing, although it must be remembered what is defined as 'high production volume' varies considerably from industry to industry.

The degree of complexity and customisation of the part has a significant contribution to the production cost of a traditionally manufactured part. Greater complexity can lead to increased machining time, operator time, additional tooling, need for custom tooling, more extensive certification, increased scrap, and the need to subcontract parts that cannot be made in house. In contrast, complexity of AM parts incurs minimal additional costs and any customisation can be done digitally and without the need for additional tools, again incurring minimal cost. This leads to figure, which shows that AM is better suited for parts with high complexity and/or customisation. Customised parts also help gain market advantage and command a premium price.

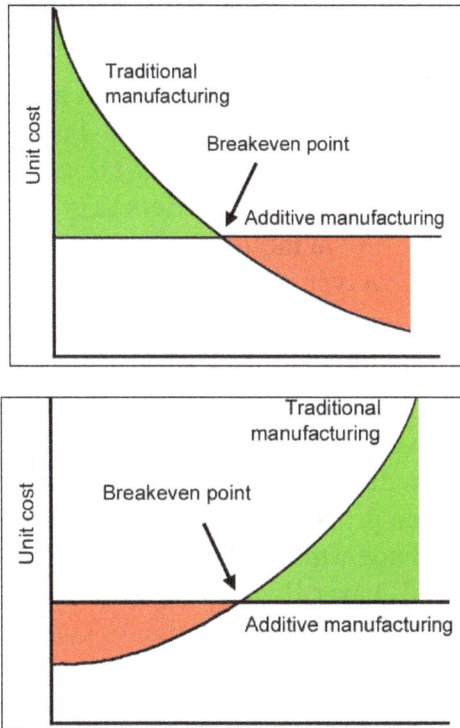

Unit production cost for additive and traditional manufacturing;
against production volume; against product complexity or customisation.

Beyond production, the effect of AM on the supply chain varies considerably from business to business, but there are many situations where it can impart savings, ether directly or by acting as an enabling technology for another process. At the beginning of 2011, the sum inventory in the manufacturing industry was $537 billion (equal to 10 % of that year's revenue). This ties up capital and building space (requiring maintenance, insurance, security etc.) and risks products deteriorating or becoming obsolete. Additive manufacturing has the potential to reduce this because, provided an inventory of suitable materials is available, it allows a wide range of parts to be produced on demand with no cost penalties for the rapid response. Another area of potential costs savings is transport costs. Additive manufacturing allows for the production of fewer, more integrated parts meaning there are simply fewer parts to transport.

An indirect benefit of AM, and potentially a very significant one, it that it is an enabling technology for remote direct digital manufacturing (DDM). Manufacturing closer to the point of sale using centrally transmitted data eliminates the need for part transport entirely and reduces the vulnerability to supply disruption for critical parts. The same on-site production method can then be used with the part (or one produced centrally) for rapid on-site production of replacement components. Although these factor can be overriding in same cases, for example use of a 3D printer in the International Space Station, analyses have indicated that in most circumstances AM machines are too specialised and expensive for this model to be cost effective at present.

Barriers to Additive Manufacturing

Additive manufacturing as it is known today has existed for approximately 30 years and it represents a limited number of industries. Current AM sales are still small compared with the machine tools industry, which had a global production value of around 60 billion in 2014, and represent only about 0.01% of the $20.5 trillion (nominal) in global manufacturing value added .

There are countless equity research organisations forecasting future markets and no consensus as to the exact scale of AM grown, but the current $2.2 billion market value is generally predicted a compound annual growth rate (CAGR) between 20% and 45%. This means AM will exceed the current machine tool market by either 2023 (45% growth) or 2030 (20% growth), but foreseeing the future is difficult and it is worth looking at the barriers to growth of AM. Gausemeier compiled a table of success factors for AM in different industries, Some weaknesses are specific to individual processes, but many are common to all types.

Surface quality, layer thickness, the ability to combine different manufacturing technologies, and the need to increase process repeatability and part reproducibility are all considered important 'process & equipment' factors. Surface quality limitations can be seen as inherent to the layered manufacturing process (partially overlapping convex tracks on an upper surface plus the so called 'step effect') although methods to reduce this such as 'intelligent' slicing, process planning and a curved layer LOM process have been proposed. It is always possible to post-process a part and to directly manufacture to near net shape with SLM and SLA by reducing layer thickness, but both these at the expense of build speed. Layer thickness is seen as important; currently they are of the order 20-100 um for SLS and SLM and 130-380 um for commercial DMD equipment. In the future, more powerful lasers may contribute to this need, but increasing energy input leads to greater risk of surface vaporization and poorer surface quality, so greater power is equally likely to facilitate increase mass deposition rates via faster laying of tracks as via laying of thicker tracks.

The furniture, jewellery, food, sports and textiles industries are taken as representative of consumer industries.

Combining different technologies will be easier to implement in some cases than others. Optomec, Aurora and Stratsasys have previously demonstrated combined operation of 3D printing and Aerosol Jet to make a "smart wing" for a small drone, demonstrating the compatibility of these technologies. This field is not well developed, but it is a need that could maybe be alleviated by combining different materials within the same process, and this is already possible in some industries. For example, manufacture of multicolour structures via FDM and manufacture of graded metallic structures or metal-matrix composites (MMCs) via DMD has been demonstrated.

The need for new materials is a factor in every sector. The current range of AM materials for all processes (atomised metal powders, proprietary polymeric powders, liquid photopolymers etc.) is restricted and much more expensive than traditional material feedstock. Nevertheless, increasing use of additive manufacturing may mean the cost of the raw materials reduces through economies of scale, reuse or standardisation of certain materials.

The need for quality assurance systems and design rules are also identified as widely important. The current lack of technical standards in AM slowing down it use for critical parts which may require long certification periods. The use of AM and DDM together also raise many questions. The need for a suitable regulatory framework in order to prevent the circulation of harmful technologies in digitalised form is very clear now that a US company, Solid Concepts, is legally selling guns and gun parts produced via laser sintering. Additive manufacturing and DDM also have the potential to blur the lines between manufacturer, wholesaler and retailer, in which case matters like legal responsibility in the case of product failure need to be resolved. Existing patent law has also been identified as barrier to the future success of additive manufacturing.

The Role of Lasers in the Future of Additive Manufacturing

The major users of additive manufacturing are not expected to change significantly in the future mid- term and have been identified as consumer products, direct medical components, transportation (automobile and aerospace), and tool and mould manufacturing. Unless a wider range of industries can be incentivised to use AM in the future, the prediction that AM will change the way companies interact and global supply chains operate on a global scale will not be accurate. Wholesale change from traditional to additive manufacturing requires a company's organisation and indeed whole culture to adapt in order to make it successful and analysis by Gilbert points to AM having difficulty penetrating high threat markets, where competition is high. Both resource rigidity (failure to change resource investment patterns) and routine rigidity (failure to change organizational processes that use those resources) are potential barriers.

Considering these major users of the technology, the trends in them and potential AM technologies are summarised in Table. It is very clear that AM is capable of tackling major goals and trends in the industries and further that many, indeed most, of these processes are laser-based.

To grow further, and become the dominant technology that many predict it to be, AM will also need to break into new industries and the above analysis does not seem to provide many incentives for that, although some of the factor in table may also apply to other industries (for example 'Accelerated product development' is becoming increasingly important in all markets and 'ageing population' is likely to affect demand for many products). However, there are at least two other factors or opportunities that could incentivise wider take up of the technology:

Table: Trends in major additive manufacturing (AM) markets and the involvement of lasers.

Market	Trends / Goals	Beneficial AM Processes	Lasers
Aerospace	Demand for lightweight structures	DMD, SLM, SLS	•
	Organic features	DMD, SLM, SLS	•
	Interior customisation	FDM, 3DP	-
	Fuel reduction	DMD, SLM, SLS	•
	Rapid tooling, fixturing	DMD, SLS	•
Automotive	Demand for lightweight structures	DMD, SLM, SLS	•
	Power train electrification	-	-
	Sustainable mobility	-	-
	Customisation	FDM, 3DP	-

Medical (dental, implants)	Increasing demand (ageing population)		-	-
	Minimally invasive surgery	SMS, SLM	●
	Replication of anatomic structures	SLS, SLA, FDM, 3DP, FDM	●
	Biomaterial manufacture	SLS, SLA, FDM/PBE	-
Electronic	Accelerated product development	Micro-SLS, SLA, AJ, 3DP	●
	Embedded electronics	3DP, AJ	-
	Miniaturization	Micro-SLS, SLA, AJ	●
	Smart microsystems	Micro-SLS, SLA, AJ	●

The Need to Reduce Environmental Impact

There is increasing emphasis on the environmental impact of manufacturing and the future is likely to see increased pressures on energy and material consumption and greenhouse gas emissions. This has the potential to become a driver for AM in the future because studies have shown that in most cases where it is economic to apply AM, it is more efficient in terms of virginal material consumption and water usage, produces less pollution and requires less landfill.

Taking Aviation as an example: It is a major contributor to greenhouse gas emissions. In 2014 greenhouse gas emissions from aviation were 688m tonnes or 2% of all human sources, and increasing globalisations means aircraft fuel use is projected to triple by 2050. Thousands of tonnes of Aluminium Titanium and Nickel alloys and greenhouse gas emission reductions of between 92.1 and 215 million metric tons during production could potentially be saved by use of AM. However, potentially even greater benefit may come from AM's ability to produce components with lighter weight, longer life or better recyclability. In independent studies, authors have identified aerospace fuel saving of 9–33% by use of AM, with consequential environmental impact, and reduction of CO_2 emissions over the whole lifecycle of a part by nearly 40% via weight saving using DMLS. If similar impact could be achieved in other industries that would be a major incentive for greater uptake of the technology.

To fully benefit from these savings, however, some advances in the AM processes are required. The amount of energy consumed when producing the same part by AM can be greater than when using conventional manufacturing processes, often due to the longer process times. Studies have found that SLS and 3D printing are less energy efficient than injection moulding at all but low production volumes where the energy to produce the mould dominates. Further, there are still potential occupation hazard, such as eye and skin irritation from metal and polymeric powders, and the difficulty in disposing of waste polymers such as epoxy resins, polycarbonates, nylons (polyamides) acrylates and styrenes, which have poor biodegradability has not been tackled.

Burkhart and Aurich proposed a tentative framework for assessing environmental effect with respect to the automotive industry, Luo a 'life stage' model for assessing the environmental effect of AM processes, and Le Bourhis a predictive model of the environmental impact to aid designers. There is work to be done, but models of this type, plus attention to the current known environmental weaknesses of AM, may be able to promote it – financially and morally - as a 'green' choice for the future.

Opportunities for Integration

Considering the range of manufacturing production and the multiple factors that affect a part's suitability for AM, it is unlikely to make traditional manufacturing processes obsolete or produce global architectural change in marketing. This points to a future with manufacturing centralised to at least some degree, and including forming, subtractive and additive techniques together. As such, the success factor "Combination of Manufacturing Technologies" that has been identified for all industry sectors (table) seems to be particularly significant, with the manufacturing technologies described extending beyond purely Additive. Integration of additive and subtractive technologies does currently exists in isolated areas. For example Microfabrica's EFAB process is a hybrid (additive/subtractive) process based on multilayer electrodeposition and planarization of at least two metals and Karunakaran proposed a weld deposition system capable of being retrofitted to a CNC machining centre. However, it is not something that is well developed at present.

Duflou described the use of multi-machine resource and energy flow ecosystems within a factory, recycling energetic or physical flows within the process chain, and between process chains. If traditional and AM systems could operate together in this way it would also have a favourable environmental impact. A decision-making framework or model to assist in selecting the manufacturing method and machine to apply to a component based on size, material, geometric features and production factors would be highly beneficial for such integrated systems .

The benefit of AM operating with traditional manufacturing as well as an independent industry become clear from the current scale of the two industries. If AM could make even a 1% increase in current global manufacturing profitability that would surpass the predicted output from the global AM industry alone more than 1000 times over. This integration could also offer an excellent opportunity for lasers, which are known for their ability to remove materials. Coupling this with a laser-based AM process greatly increases the range of parts that can be produced and brings the prospect of the laser shopfloor closer.

Electroforming

Electroforming is an additive manufacturing process specialized for the production of high precision metal parts. Its uniqueness is that you can grow metal parts atom by atom, providing extreme accuracy and high aspect ratios.

Once you've discovered the benefits of Electroforming, a whole new world of opportunities opens up. What if you could produce, atom by atom, stress- and burr-free precision metal parts with micron scale accuracy? It would give you the opportunity to raise the bar on precision, tolerance, cost-effectiveness and the capability to withstand higher temperatures.

The Electroforming Process

While known as a highly accurate additive manufacturing process, Electroforming is also an electro-deposition process.

The Electroforming process can be concluded in a series of steps including Cleaning, Coating, Exposing, Developing, Deposition, and Harvesting.

The electroforming process allows extreme precise duplication of the mandrel. The high resolution of the conductive patterned substrate allows finer geometries, tighter tolerances and superior edge definition. This results in perfect process control, high quality production and very high repeatability. This makes electroforming perfect suitable for low cost production and high volumes.

Cleaning: The metal substrate will first be cleaned and degreased.

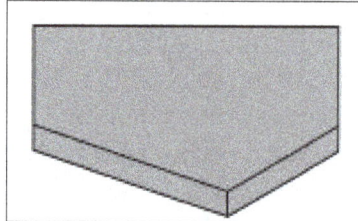

Coating: The cleaned metal 'blank' is then coated with a light-sensitive coating/photoresist.

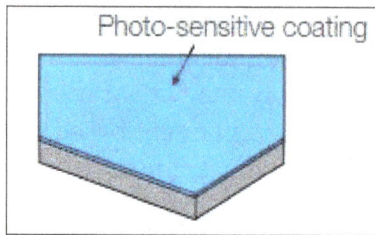

Exposing: The metal sheet is then exposed to ultra-violet light, which hardens the photoresist.

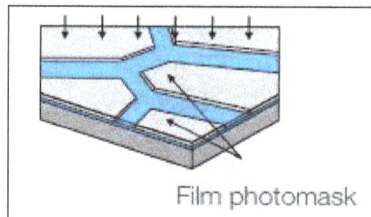

Developing: After electrolytic bath is used to deposit metal onto the patterned surface.

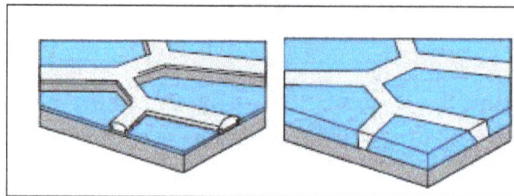

Harvesting: The electroformed part can be harvested from the mandrel, once the material is plated in the desired thickness.

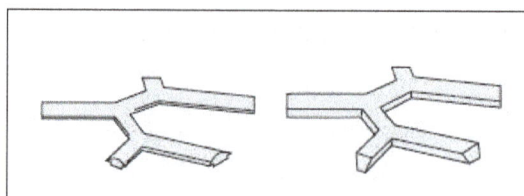

References

- Dohda, Kuniaki; Ni, Jun (November 2004). "Micro/Meso-scale Manufacturing". Journal of Manufacturing Science and Engineering

- Manufacturing-process-meaning-and-types: engineeringarticles.org, Retrieved 15 April, 2019

- Thoury, M.; et al. (2016). "High spatial dynamics-photoluminescence imaging reveals the metallurgy of the earliest lost-wax cast object". Nature Communications. 7. Doi:10.1038/ncomms13356

- Laser-engraving-definition, glossary, en: sculpteo.com, Retrieved 16 May, 2019

- Degarmo, E. Paul; Black, J T.; Kohser, Ronald A. (2003), Materials and Processes in Manufacturing (9th ed.), Wiley, p. 277, ISBN 0-471-65653-4

- Plastics-moulding-methods-dip, plastics-moulding-methods: rutlandplastics.co.uk, Retrieved 17 June, 2019

- Donaldson, Brent (2017-11-01), "Foundry Says Robotic Sand Printing a "Game Changer" for Metal Casting", Additive Manufacturing, retrieved 2017-11-14

- Metal-injection-molding, wu: custompartnet.com, Retrieved 18 July, 2019

- Jan Schroers; Thomas M. Hodges; Golden Kumar; Hari Raman; Anthony J. Barnes; Quoc Pham; Theodore A. Waniuk (February 2011). "Thermoplastic blow molding of metals". Materials Today. 14: 14–19. Doi:10.1016/S1369-7021(11)70018-9

- Electroforming, technologies: vecoprecision.com, Retrieved 19 August, 2019

- Martin, Alan; Harbison, Sam; Beach, Karen; Cole, Peter (2012-03-30). An Introduction to Radiation Protection 6E. CRC Press. ISBN 9781444146073

Manufacturing Tools and Technologies

Manufacturing engineering requires numerous tools and technologies to perform various operations. Conveyor system, shaper, optical comparator, water jet cutter, lathe, jig, pattern, fixture, die, material-handling tools, etc. are some of the tools and technologies used. This chapter closely examines these tools and technologies used in manufacturing engineering to provide an extensive understanding of the subject.

MANUFACTURING TECHNOLOGY

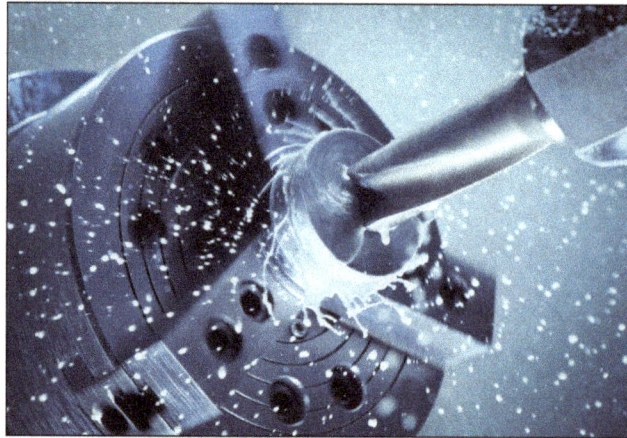

Manufacturing technology provides the tools that enable production of all manufactured goods. These master tools of industry magnify the effort of individual workers and give an industrial nation the power to turn raw materials into the affordable, quality goods essential to today's society.

Manufacturing technology provides the productive tools that power a growing, stable economy and a rising standard of living. These tools create the means to provide an effective national defense. They make possible modern communications, affordable agricultural products, efficient transportation, innovative medical procedures, space exploration and the everyday conveniences we take for granted.

Production tools include machine tools and other related equipment and their accessories and tooling. Machine tools are non-portable, power-driven manufacturing machinery and systems used to perform specific operations on man-made materials to produce durable goods or components. Related technologies include Computer Aided Design (CAD) and Computer Aided Manufacturing (CAM) as well as assembly and test systems to create a final product or subassembly.

Today's increasingly automated and software driven industries have reduced human intervention to pressing only a few buttons in some cases. The application of advanced technologies in manufacturing such as nanotechnology, cloud computing, the Internet of Things (IoT) are changing the face of manufacturing in ways unimaginable a few decades ago. In addition to cutting the costs, these technologies create speed, precision, efficiency and flexibility for manufacturing companies.

3D Printing

One of the biggest news in the manufacturing technology sector in the last few years is the proliferation and application of 3D printing technology. It has caught the imagination of the general public and the manufacturing community like nothing since the invention of the personal computer and the internet. Within a few years, the technology has evolved so much that it is now possible to produce almost any component using metal, plastic, mixed materials and even human tissue. It has forced engineers and designers to think very differently when thinking about product development. As this more manufacturers adopt and use 3D printing technology, there is little doubt that 3D Printing will change the face of manufacturing forever.

Nanotechnology

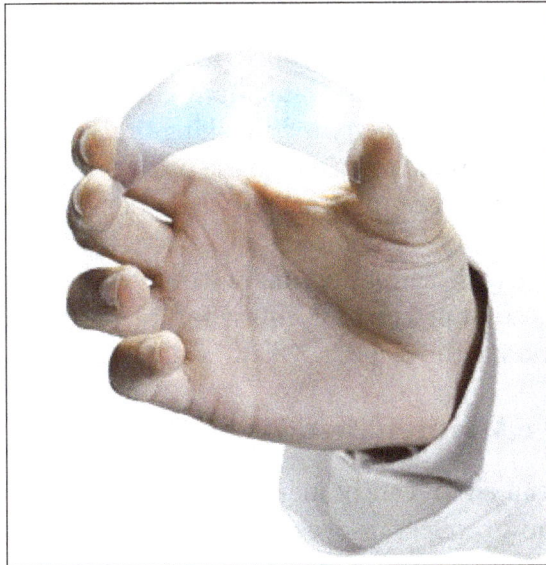

Nanotechnology is the technology but the first generation of the technology is already here. It involves the manipulation of matter on atomic, molecular and supramolecular scales; thus bringing with it super-precision manufacturing. Currently applied mostly in space technology and biotechnology, it is going to play an indispensable role in every manufacturing industry. In many ways, it has already changed the world. Examples of application in nanotechnology include:

- Faster computer processing,

- Smaller memory cards that have more memory space,

- Clothes that last longer and keep the wearer cool in the summer,

- Bandages that heal wounds faster,

- And tennis and bowling balls that last longer.

There will be nanobots (microscopic robots) that will carry drugs to specific tissues in our body.

The Internet of Things (IoT)

The Internet of Things (IoT) is a revolutionary manufacturing technology that allows electronic devices connected to each other, within the existing Internet infrastructure, to communicate with one another without human intervention. An IoT device connects to the internet and is capable of generating and receiving signals. As such, the use of this technology is going to have a profound impact on the manufacturing industry. IoT enables connected devices to talk to each other, sending and receiving critical notifications. An example of a critical notification is a defect or damaged ping. Once the device detects a failure, the IoT connected device sends a notification to another device or a user. This type of small, but critical, application of IoT in manufacturing results in reduced downtime, increased quality, reduced waste and less overall costs.

Cloud Computing

Cloud computing is the practice of using a network of Internet-connected remote services along various points to store, manage, and process data. Many companies are already using cloud computing, although the manufacturing industry is still taking time to warm up to the technology due to connectivity and security concerns. Over time, to the present day, cloud computing grows more stable and reliable. Manufacturers are increasingly implementing cloud computing software in manufacturing plants spread out in various geographic areas in order to share data quickly and efficiently. In implementing cloud computing, manufacturers reduce costs, gain greater quality control, and increase the speed of production. In the future, it is feasible that all manufacturing facilities will have a connection to the cloud.

Big Data and Predictive Maintenance Technology

Manufacturing industries can significantly increase their efficiency and productivity with the technologies that allow them to collect, process and measure big data in real time. These technologies include electronic devices that connect factories through the internet and web pages that double as dashboards for controlling the processes. Predictive maintenance technology helps predict snags and defects and thus cuts downtime and costs. In the future, manufacturers will implement big data and predictive maintenance technologies in every area of manufacturing. IoT is a part of big data and predictive technology that manufacturers are already using with remarkable success.

Advanced technologies have been the driving force behind the growth of the manufacturing industries, and they will have a greater role to play in the industries. As new technologies emerge, manufacturers will adopt them, or they will be forced to choose them to survive. On their part, the technologies will change the industries beyond recognition. For example, 3D printing is already changing the way many manufacturers design and manufacture their products.

FLEXIBLE MANUFACTURING SYSTEM

A flexible manufacturing system (FMS) is a manufacturing system in which there is some amount of flexibility that allows the system to react in case of changes, whether predicted or unpredicted. This flexibility is generally considered to fall into two categories, which both contain numerous subcategories.

The first category, routing flexibility, covers the system's ability to be changed to produce new product types and ability to change the order of operations executed on a part. The second category is called machine flexibility, which consists of the ability to use multiple machines to perform the same operation on a part, as well as the system's ability to absorb large-scale changes, such as in volume, capacity, or capability.

Most FMS consist of three main systems. The work machines which are often automated CNC machines are connected by a material handling system to optimize parts flow and the central control computer which controls material movements and machine flow.

The main advantages of an FMS is its high flexibility in managing manufacturing resources like time and effort in order to manufacture a new product. The best application of an FMS is found in the production of small sets of products like those from a mass production.

Advantages

- Reduced manufacturing cost,
- Lower cost per unit produced,
- Greater labor productivity,
- Greater machine efficiency,
- Improved quality,
- Increased system reliability,
- Reduced parts inventories,
- Adaptability to CAD/CAM operations,
- Shorter lead times,
- Improved efficiency,
- Increase production rate,
- Disadvantages,
- Initial set-up cost is high,
- Substantial pre-planning,
- Requirement of skilled labor,

- Complicated system,

- Maintenance is complicated.

Flexibility

Flexibility in manufacturing means the ability to deal with slightly or greatly mixed parts, to allow variation in parts assembly and variations in process sequence, change the production volume and change the design of certain product being manufactured.

Industrial FMS Communication

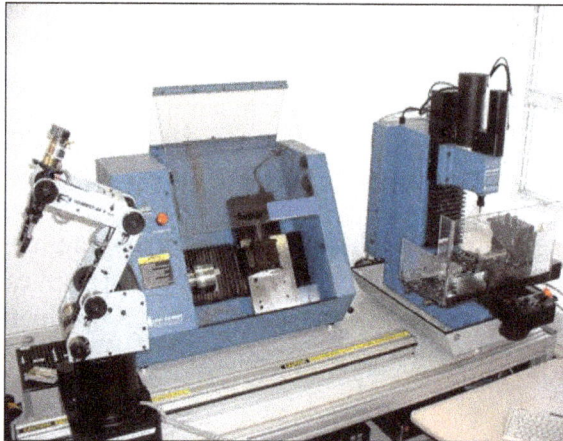

Training FMS with learning robot <u>SCORBOT-ER 4u</u>, workbench CNC Mill and CNC Lathe.

An Industrial Flexible Manufacturing System (FMS) consists of robots, Computer-controlled Machines, Computer Numerical Controlled machines (CNC), instrumentation devices, computers, sensors, and other stand alone systems such as inspection machines. The use of robots in the production segment of manufacturing industries promises a variety of benefits ranging from high utilization to high volume of productivity. Each Robotic cell or node will be located along a material handling system such as a conveyor or automatic guided vehicle. The production of each part or work-piece will require a different combination of manufacturing nodes. The movement of parts from one node to another is done through the material handling system. At the end of part processing, the finished parts will be routed to an automatic inspection node, and subsequently unloaded from the Flexible Manufacturing System.

The FMS data traffic consists of large files and short messages, and mostly come from nodes, devices and instruments. The message size ranges between a few bytes to several hundreds of bytes. Executive software and other data, for example, are files with a large size, while messages for machining data, instrument to instrument communications, status monitoring, and data reporting are transmitted in small size.

There is also some variation on response time. Large program files from a main computer usually take about 60 seconds to be down loaded into each instrument or node at the beginning of FMS operation. Messages for instrument data need to be sent in a periodic time with deterministic time delay. Other types of messages used for emergency reporting are quite short in size and must be transmitted and received with an almost instantaneous response. The demands for reliable FMS protocol that support all the FMS data characteristics are now urgent. The existing IEEE standard protocols do not fully satisfy the real time communication requirements in this environment. The delay of CSMA/CD is unbounded as the number of nodes increases due to the message collisions. Token Bus has a deterministic message delay, but it does not support prioritized access scheme which is needed in FMS communications. Token Ring provides prioritized access and has a low message delay, however, its data transmission is unreliable. A single node failure which may occur quite often in FMS causes transmission errors of passing message in that node. In addition, the topology of Token Ring results in high wiring installation and cost.

CNC machine.

A design of FMS communication that supports a real time communication with bounded message delay and reacts promptly to any emergency signal is needed. Because of machine failure and malfunction due to heat, dust, and electromagnetic interference is common, a prioritized mechanism and immediate transmission of emergency messages are needed so that a suitable recovery procedure can be applied. A modification of standard Token Bus to implement a prioritized access scheme was proposed to allow transmission of short and periodic messages with a low delay compared to the one for long messages.

CONVEYOR SYSTEM

An overhead chain conveyor conveys cars at Mercedes in Germany.

A conveyor system is a common piece of mechanical handling equipment that moves materials from one location to another. Conveyors are especially useful in applications involving the transportation of heavy or bulky materials. Conveyor systems allow quick and efficient transportation for a wide variety of materials, which make them very popular in the material handling and packaging industries. They also have popular consumer applications, as they are often found in supermarkets and airports, constituting the final leg of item/ bag delivery to customers. Many kinds of conveying systems are available and are used according to the various needs of different industries. There are chain conveyors (floor and overhead) as well. Chain conveyors consist of enclosed tracks, I-Beam, towline, power & free, and hand pushed trolleys.

Industries that use Conveyor Systems

A lineshaft roller conveyor conveys boxed produce at a distribution center.

A Conveyor belt conveys papers at a newspaper print plant.

Roller conveyor for carton transport in the apparel industry.

Conveyor systems are used widespread across a range of industries due to the numerous benefits they provide.

- Conveyors are able to safely transport materials from one level to another, which when done by human labor would be strenuous and expensive.

- They can be installed almost anywhere, and are much safer than using a forklift or other machine to move materials.

- They can move loads of all shapes, sizes and weights. Also, many have advanced safety features that help prevent accidents.

- There are a variety of options available for running conveying systems, including the hydraulic, mechanical and fully automated systems, which are equipped to fit individual needs.

Conveyor systems are commonly used in many industries, including the mining, automotive, agricultural, computer, electronic, food processing, aerospace, pharmaceutical, chemical, bottling and canning, print finishing and packaging. Although a wide variety of materials can be conveyed, some of the most common include food items such as beans and nuts, bottles and cans, automotive components, scrap metal, pills and powders, wood and furniture and grain and animal feed. Many factors are important in the accurate selection of a conveyor system. It is important to know how the conveyor system will be used beforehand. Some individual areas that are helpful to consider are the required conveyor operations, such as transportation, accumulation and sorting, the material sizes, weights and shapes and where the loading and pickup points need to be.

Care and Maintenance of Conveyor Systems

A conveyor system is often the lifeline to a company's ability to effectively move its product in a timely fashion. The steps that a company can take to ensure that it performs at peak capacity, include regular inspections and system audits, close monitoring of motors and reducers, keeping key parts in stock, and proper training of personnel.

Increasing the service life of a conveyor system involves: choosing the right conveyor type, the right system design and paying attention to regular maintenance practices.

A conveyor system that is designed properly will last a long time with proper maintenance. Here are six of the biggest problems to watch for in overhead type conveyor systems including I-beam monorails, enclosed track conveyors and power and free conveyors. Overhead conveyor systems have been used in numerous applications from shop displays, assembly lines to paint finishing plants and more.

Poor take-up adjustment: This is a simple adjustment on most systems yet it is often overlooked. The chain take-up device ensures that the chain is pulled tight as it leaves the drive unit. As wear occurs and the chain lengthens, the take-up extends under the force of its springs. As they extend, the spring force becomes less and the take-up has less effect. Simply compress the take-up springs and your problem goes away. Failure to do this can result in chain surging, jamming, and extreme wear on the track and chain. Take-up adjustment is also important for any conveyor using belts as a means to power rollers, or belts themselves being the mover. With poor-take up on belt-driven rollers, the belt may twist into the drive unit and cause damage, or at the least a noticeable decrease or complete loss of performance may occur. In the case of belt conveyors, a poor take-up may cause drive unit damage or may let the belt slip off of the side of the chassis.

Lack of lubrication: Chain bearings require lubrication in order to reduce friction. The chain pull that the drive experiences can double if the bearings are not lubricated. This can cause the system to overload by either its mechanical or electrical overload protection. On conveyors that go through hot ovens, lubricators can be left on constantly or set to turn on every few cycles.

Contamination: Paint, powder, acid or alkaline fluids, abrasives, glass bead, steel shot, etc. can all lead to rapid deterioration of track and chain. Ask any bearing company about the leading cause of bearing failure and they will point to contamination. Once a foreign substance lands on the raceway of a bearing or on the track, pitting of the surface will occur, and once the surface is compromised, wear will accelerate. Building shrouds around your conveyors can help prevent the ingress of contaminants. Or, pressurize the contained area using a simple fan and duct arrangement. Contamination can also apply to belts (causing slippage, or in the case of some materials premature wear), and of the motors themselves. Since the motors can generate a considerable amount of heat, keeping the surface clean is an almost-free maintenance procedure that can keep heat from getting trapped by dust and grime, which may lead to motor burnout.

Product handling: In conveyor systems that may be suited for a wide variety of products, such as those in distribution centers, it is important that each new product be deemed acceptable for conveying before being run through the materials handling equipment. Boxes that are too small, too large, too heavy, too light, or too awkwardly shaped may not convey, or may cause many problems including jams, excess wear on conveying equipment, motor overloads, belt breakage, or other damage, and may also consume extra man-hours in terms of picking up cases that slipped between rollers, or damaged product that was not meant for materials handling. If a product such as this manages to make it through most of the system, the sortation system will most likely be the affected, causing jams and failing to properly place items where they are assigned. Any and all cartons handled on any conveyor should be in good shape or spills, jams, downtime, and possible accidents and injuries may result.

Drive train: Notwithstanding the above, involving take-up adjustment, other parts of the drive train should be kept in proper shape. Broken O-rings on a Lineshaft, pneumatic parts in disrepair, and motor reducers should also be inspected. Loss of power to even one or a few rollers on a conveyor can mean the difference between effective and timely delivery, and repetitive nuances that can continually cost downtime.

Bad belt tracking or timing: In a system that uses precisely controlled belts, such as a sorter system, regular inspections should be made that all belts are traveling at the proper speeds at all times. While usually a computer controls this with Pulse Position Indicators, any belt not controlled must be monitored to ensure accuracy and reduce the likelihood of problems. Timing is also important for any equipment that is instructed to precisely meter out items, such as a merge where one box pulls from all lines at one time. If one were to be mistimed, product would collide and disrupt operation. Timing is also important wherever a conveyor must "keep track" of where a box is, or improper operation will result.

Since a conveyor system is a critical link in a company's ability to move its products in a timely fashion, any disruption of its operation can be costly. Most downtime can be avoided by taking steps to ensure a system operates at peak performance, including regular inspections, close monitoring of motors and reducers, keeping key parts in stock, and proper training of personnel.

Impact and Wear-resistant Materials used in Conveyor Systems

Conveyor systems require materials suited to the displacement of heavy loads and the wear-resistance to hold-up over time without seizing due to deformation. Where static control is a factor, special materials designed to either dissipate or conduct electrical charges are used. Examples of conveyor handling materials include UHMW, nylon, Nylatron NSM, HDPE, Tivar, Tivar ESd, and polyurethane.

Growth of Conveyor Systems in Various Industries

As far as growth is concerned the material handling and conveyor system makers are getting utmost exposure in the industries like automotive, pharmaceutical, packaging and different production plants. The portable conveyors are likewise growing fast in the construction sector and by the year 2014 the purchase rate for conveyor systems in North America, Europe and Asia is likely to grow even further. The most commonly purchased types of conveyors are Line shaft roller conveyor, chain conveyors and conveyor belts at packaging factories and industrial plants where usually product finishing and monitoring are carried. Commercial and civil sectors are increasingly implementing conveyors at airports, shopping malls, etc.

Types of Conveyor Systems

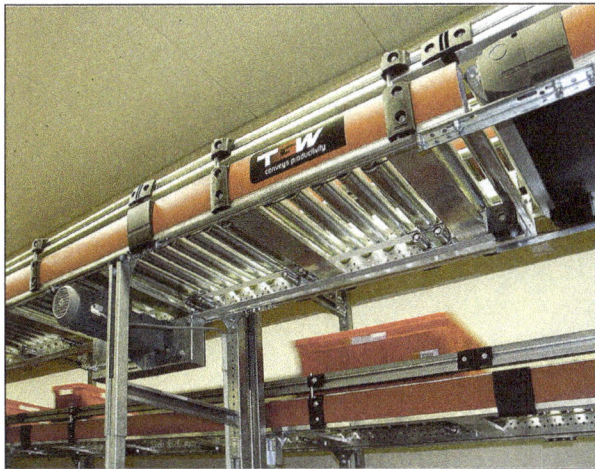

Belt driven roller conveyor for cartons and totes.

Flexible conveyor.

- Aero-mechanical conveyors,
- Automotive conveyors,
- Belt conveyor,
- Belt-driven live roller conveyors,
- Bucket conveyor,
- Chain conveyor,
- Chain-driven live roller conveyor,
- Drag conveyor,
- Dust-proof conveyors,
- Electric track vehicle systems,
- Flexible conveyors,
- Gravity conveyor,
- Gravity skatewheel conveyor,
- Lineshaft roller conveyor,
- Motorized-drive roller conveyor,
- Overhead I-beam conveyors,
- Overland conveyor,
- Pharmaceutical conveyors,
- Plastic belt conveyors,
- Pneumatic conveyors,
- Screw or auger conveyor,
- Spiral conveyors,
- Vertical conveyors,
- Vibrating conveyors,
- Wire mesh conveyors.

Pneumatic Conveyor Systems

Every pneumatic system uses pipes or ducts called transportation lines that carry a mixture of materials and a stream of air. These materials are free flowing powdery materials like cement and fly ash. Products are moved through tubes by air pressure. Pneumatic conveyors are either carrier systems or dilute-phase systems; carrier systems simply push items from one entry point to one exit point, such as the money-exchanging pneumatic tubes used at a bank drive-through window. Dilute-phase systems use push-pull pressure to guide materials through various entry and exit points. Air compressors or blowers can be used to generate the air flow. Three systems used to generate high-velocity air stream:

- Suction or vacuum systems, utilizing a vacuum created in the pipeline to draw the material with the surrounding air. The system operated at a low pressure, which is practically 0.4–0.5 atm below atmosphere, and is utilized mainly in conveying light free flowing materials.

- Pressure-type systems, in which a positive pressure is used to push material from one point to the next. The system is ideal for conveying material from one loading point to a number of unloading points. It operates at a pressure of 6 atm and upwards.

- Combination systems, in which a suction system is used to convey material from a number of loading points and a pressure system is employed to deliver it to a number of unloading points.

Vibrating Conveyor Systems

A vibrating conveyor is a machine with a solid conveying surface which is turned up on the side to form a trough. They are used extensively in food-grade applications to convey dry bulk solids where sanitation, washdown, and low maintenance are essential. Vibrating conveyors are also suitable for harsh, very hot, dirty, or corrosive environments. They can be used to convey newly-cast metal parts which may reach upwards of 1,500 °F (820 °C). Due to the fixed nature of the conveying pans vibrating conveyors can also perform tasks such as sorting, screening, classifying and orienting parts. Vibrating conveyors have been built to convey material at angles exceeding 45° from horizontal using special pan shapes. Flat pans will convey most materials at a 5° incline from horizontal line.

Flexible Conveyor Systems

The flexible conveyor is based on a conveyor beam in aluminum or stainless steel, with low-friction slide rails guiding a plastic multi-flexing chain. Products to be conveyed travel directly on the conveyor, or on pallets/carriers. These conveyors can be worked around obstacles and keep production lines flowing. They are made at varying levels and can work in multiple environments. They are used in food packaging, case packing, and pharmaceutical industries and also in large retail stores such as Wal-Mart and Kmart.

Vertical Conveyor Systems and Spiral Conveyors

Vertical conveyors, also commonly referred to as freight lifts and material lifts, are conveyor systems used to raise or lower materials to different levels of a facility during the handling process. Examples of these conveyors applied in the industrial assembly process include transporting materials to different floors. While similar in look to freight elevators, vertical conveyors are not equipped to transport people, only materials.

Vertical lift conveyors contain two adjacent, parallel conveyors for simultaneous upward movement of adjacent surfaces of the parallel conveyors. One of the conveyors normally has spaced apart flights (pans) for transporting bulk food items. The dual conveyors rotate in opposite directions, but are operated from one gear box to ensure equal belt speed. One of the conveyors is pivotally hinged to the other conveyor for swinging the attached conveyor away from the remaining conveyor for access to the facing surfaces of the parallel conveyors. Vertical lift conveyors can be manually or automatically loaded and controlled. Almost all vertical conveyors can be systematically integrated with horizontal conveyors, since both of these conveyor systems work in tandem to create a cohesive material handling assembly line.

Like vertical conveyors, spiral conveyors raise and lower materials to different levels of a facility. In contrast, spiral conveyors are able to transport material loads in a continuous flow. A helical spiral or screw rotates within a sealed tube and the speed makes the product in the conveyor rotate with the screw. The tumbling effect provides a homogeneous mix of particles in the conveyor, which is essential when feeding pre-mixed ingredients and maintaining mixing integrity. Industries that require a higher output of materials - food and beverage, retail case packaging, pharmaceuticals typically incorporate these conveyors into their systems over standard vertical conveyors due to their ability to facilitate high throughput. Most spiral conveyors also have a lower angle of incline or decline (11 degrees or less) to prevent sliding and tumbling during operation.

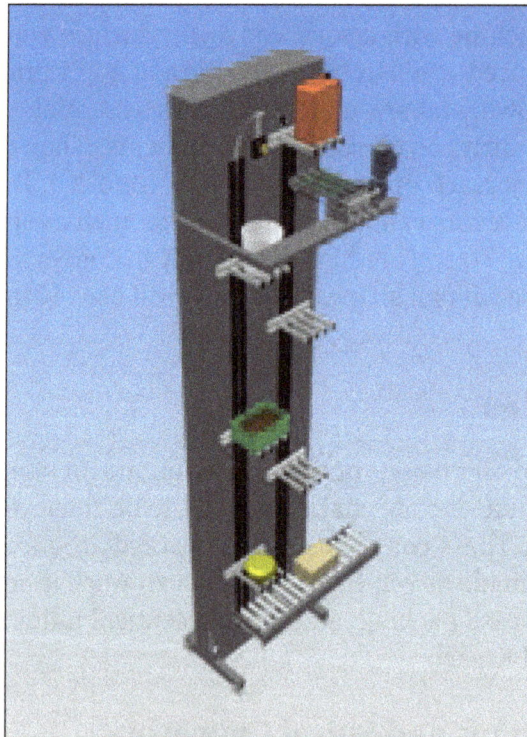

Vertical conveyor with forks.

Like spiral conveyors, vertical conveyors that use forks can transport material loads in a continuous flow. With these forks the load can be taken from one horizontal conveyor and put down on another horizontal conveyor on a different level. By adding more forks, more products can be lifted at the same time. Conventional vertical conveyors must have input and output of material loads moving in the same direction. By using forks many combinations of different input- and output- levels in different directions are possible. A vertical conveyor with forks can even be used as a vertical sorter. Compared to a spiral conveyor a vertical conveyor with or without forks - takes up less space.

Vertical reciprocating conveyors (or VRC) are another type of unit handling system. Typical applications include moving unit loads between floor levels, working with multiple accumulation conveyors, and interfacing overhead conveyors line. Common material to be conveyed includes pallets, sacks, custom fixtures or product racks and more.

Heavy-duty Roller Conveyors

Heavy-duty roller conveyors are used for moving items that weigh at least 500 pounds (230 kg). This type of conveyor makes the handling of such heavy equipment/products easier and more time effective. Many of the heavy duty roller conveyors can move as fast as 75 feet per minute (23 m/min).

Other types of heavy-duty roller conveyors are gravity roller conveyors, chain-driven live roller conveyors, pallet accumulation conveyors, multi-strand chain conveyors, and chain and roller transfers.

Gravity roller conveyors are easy to use and are used in many different types of industries such as automotive and retail.

Chain-driven live roller conveyors are used for single or bi-directional material handling. Large, heavy loads are moved by chain driven live roller conveyors.

Pallet accumulation conveyors are powered through a mechanical clutch. This is used instead of individually powered and controlled sections of conveyors.

Multi-strand chain conveyors are used for double-pitch roller chains. Products that cannot be moved on traditional roller conveyors can be moved by a multi-strand chain conveyor.

Chain and roller conveyors are short runs of two or more strands of double-pitch chain conveyors built into a chain-driven line roller conveyor. These pop up under the load and move the load off of the conveyor.

SHAPER

The shaper machine is a reciprocating type of machine basically used for producing the horizontal, vertical or flat surfaces. The shaper holds the single point cutting tool in ram and workpiece is fixed in the table.

During the forward stroke, the ram is holding the tool is reciprocating over the workpiece to cut into the required shape. During the return stroke, no metal is cutting. In the shaper machine, the rotary motion of the drive is converted into reciprocating motion of ram holding the tool.

Therefore in order to reduce the total machine time, It allows the ram holding the tool should move slower during forwarding cutting stroke and it comes faster in return stroke. This can be achieved by a mechanism called a quick return mechanism.

Shaper Machine Process

The shaper process can be defined as a process for removing metal from the surface in horizontal, vertical and angular planes by the use of a single point cutting tool held in a ram that reciprocates the tool in a linear direction across the workpiece held on the table of the machine. The work is fed at right angles to the direction of the ram in small increments, at the end of the return stroke.

Parts of Shaper Machine

The following are the main parts of shaper machine:

- Base,
- Column,
- Cross-rail,
- Table,
- Ram.

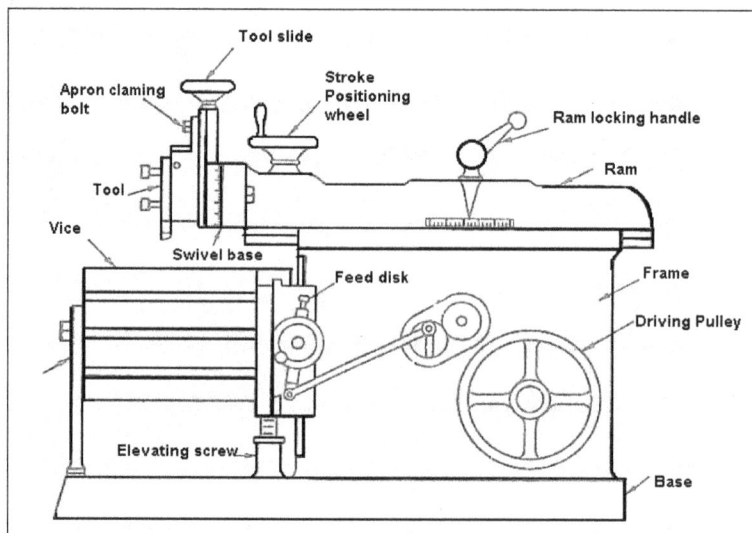

The arrangement of shaper machine is made as shown in the figure. It consists of the following parts.

Base

- The base is the necessary bed or support required for all machines tools.
- The base is hollow casting made of cast iron to resist vibration and on which all parts of the shaper are mounted.
- It is so designed that is can take up the entire load of the machine and the forces set up by cutting tool over the work.

Column

- This is made of cast iron, which is a box-like and is mounted on the base.

- Two accurately machined guideways are provided on the top of the column on which the ram reciprocates.

- The column acts as a cover to the drive mechanism and also supports the reciprocating ram and the worktable.

Cross-rail

- Cross rail is mounted on the front vertical surface of the column on which saddle is mounted.

- The vertical movement is given to the table by raising or lowering the cross rail using the elevating screw.

- The horizontal movement is given to the table by moving the saddle using the crossfeed screw.

Table

- The table is bolted to the saddle and receives crosswise and vertical movements from saddle cross rail.

- T-bolts are used for clamping on top and sides.

- The table can be swiveled at any required angle.

- In a universal shaper, the table may be swiveled on a horizontal axis and the upper part of the table may be fitted up or down.

- In heavier type shaper the table clamped with table support to make it more rigid.

Ram

- The ram reciprocates on the column guideways and carries the tool head with a single point cutting tool.

- The tool head is in the clapper box, which causes cutting action only in a forward stroke of the ram and sliding movement of the tool in the reverse stroke of the ram.

- The depth of cut or feed of the tool is given by down feed screw.

- The tool head has swivel base degree graduations, which helps to move the tool head to any desired inclination for machining inclined surfaces on the workpieces.

Types of Shaper Machines

Following are the different types of shaper machines.

- Based on the type of driving mechanism.

  o Crank type shaper.

  o Geared type shaper.

  o Hydraulic type shaper.

- Based on ram travel.
  - Horizontal shaper.
  - Vertical shaper.
- Based on the table design.
  - Standard shaper.
  - Universal shaper.
- Based on cutting stroke.
  - Push cut type.
  - Draw cut type.

Bases on the Type of Driving Mechanism

Following are the different types of shaper machines based on the type of driving mechanism.

Crank Type Shaper Machine

These are very common types of shaper machines, which is using to hold the workpiece on the table. The tool is reciprocating in motion equal to the length of the stroke desired while the work is clamped in position on an adjustable table.

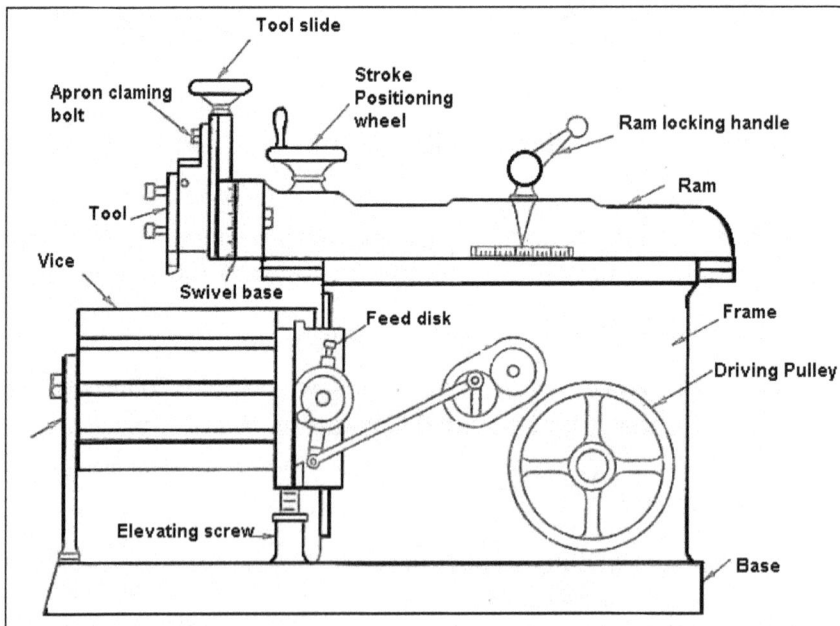

In construction, the crank shaper employs a crank mechanism to change the circular motion of a large gear called "bull gear" incorporated in the machine to reciprocation motion of the ram.

It uses a crank mechanism to convert the circular motion of the bull gear into reciprocating motion of the ram. The ram carries a tool head at its end & provides the cutting action.

Gear Type Shaper Machine

In these types of shaper machines, the ram is reciprocating. The ram is affecting due to reciprocating motion with the rack and pinion. The rack teeth are cut directly below the ram mesh with the spur gear.

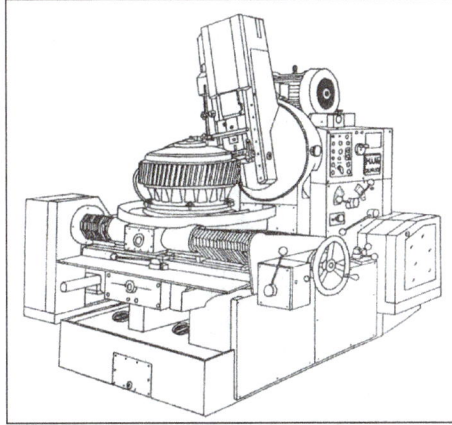

Gear Type Shaper Machine.

The speed and the direction in which the machine will traverse depend on the number of gears in the gear train. This type of shaper machines is not widely using in any industry.

Hydraulic Shaper Machine

In this types of shaper machines, the reciprocating motion of the ram is provided by the hydraulic mechanism. The Hydraulic shaper uses the oil under high pressure. The end of the piston rod is connected to the ram.

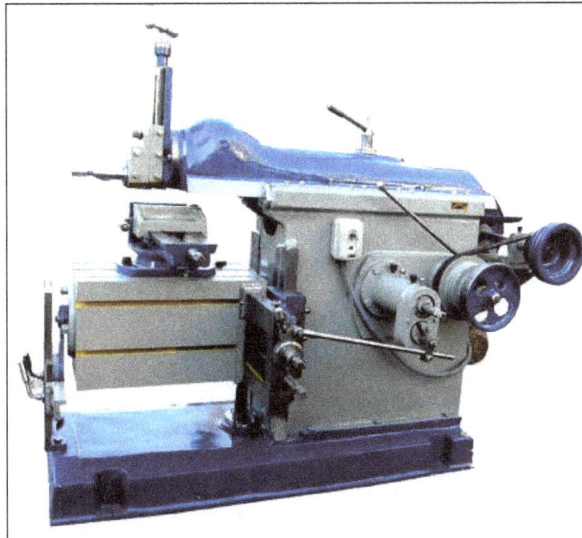

Hydraulic Type Shaper Machine.

The high-pressure oil first acts on one side of the piston and then on the other causing the piston to reciprocating and the motion is transmitted to the ram. The main advantages of this type of shaper machine are that the cutting speed and force of the ram drive are constant. From start to end of the cut without making noise and operates quietly.

Based on Ram Travel

Following are the different types of shaper machine based on ram travel.

Horizontal Shaper Machine

Horizontal Shaper Machine.

In these types of shaper machines, the ram is reciprocating. The ram holding the tool in a horizontal axis and reciprocate. This type of shaper is using for the production of flat surfaces, external grooves, keyways etc.

Vertical Shaper Machine

In these types of shaper machines, the ram reciprocating in verticle plane. In this, the table holds the workpiece. Verticle shapers maybe crank driven, rack-driven, screw-driven or hydraulic power-driven.

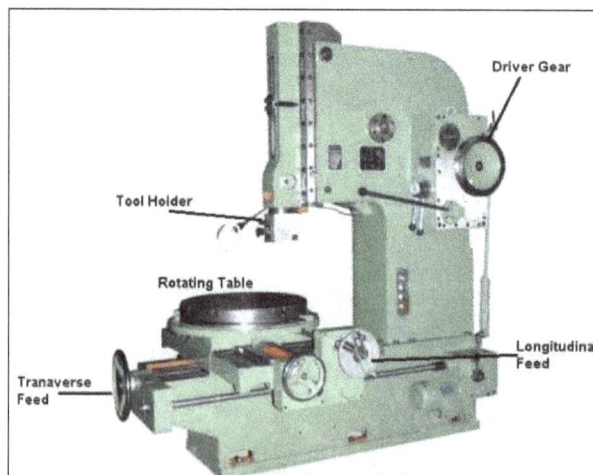

Vertical Type Shaper Machine.

The vertical shaper is very convenient for machining internal surfaces, keyways, slots or grooves. The workpiece can move in any given directions such as the cross, longitudinal or rotary movements. This type of shaper is suitable for machining internal surfaces, slots & keyways.

Based on the Table Design

Following are the different types of shaper machine based on the table design.

Standard Shaper Machine

In this types of shaper machines, the table has only two movements, vertical and horizontal, to give the feed. That's why it known as standard shaper machine. Here the table is not supporting at the outer end.

Universal Shaper Machine

In this types of shaper machines, in addition to the two moments i.e. vertical and horizontal, the table can be moving in an inclined axis and also it can swivel on its own axis.

Since the workpiece mounted on the can be adjusted in different planes, the shaper os suitable for a different type of operations and is given the name "Universal". This type of shaper is commonly using the tool room works.

Based on Cutting Stroke

Following are the different types of shaper machine based on cutting stroke.

Push cut Shaper Machine

In this types of shaper machines, the metal is removed in the forward motion of the ram. This is commonly used types of shaper machines.

Draw cut Shaper Machine

In this types of shaper machines, the metal is removed in the backward motion of the ram. In this shaper, the tool is fixed in the tool head in the reverse direction so that it provides the cutting action in the reverse stroke of the ram.

OPTICAL COMPARATOR

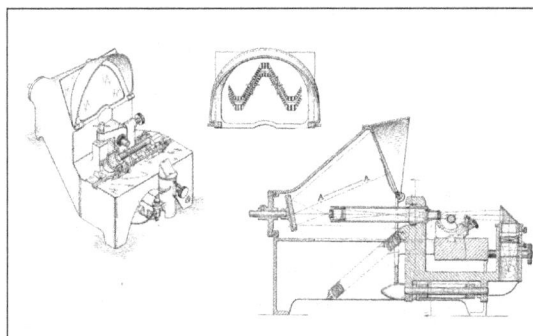

Patent drawings for Hartness screw-thread optical comparator (numbering removed for clarity).

A J&L comparator with a DRO.

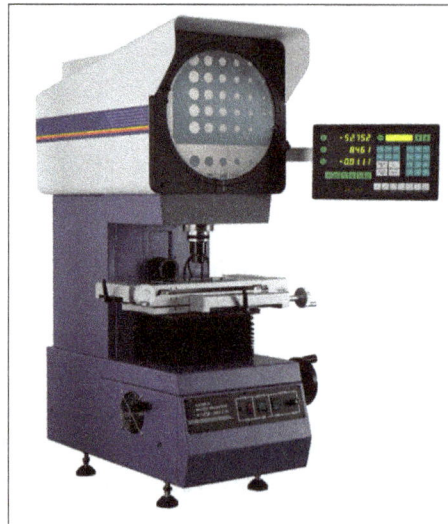

Profile projector, also known as contour comparator, is widely used to measure 2-dimensional data.

An optical comparator (often called just a comparator in context) or profile projector is a device that applies the principles of optics to the inspection of manufactured parts. In a comparator, the magnified silhouette of a part is projected upon the screen, and the dimensions and geometry of the part are measured against prescribed limits. It is a useful item in a small parts machine shop or production line for the quality control inspection team.

The measuring happens in any of several ways. The simplest way is that graduations on the screen, being superimposed over the silhouette, allow the viewer to measure, as if a clear ruler were laid over the image. Another way is that various points on the silhouette are lined up with the reticle at the centerpoint of the screen, one after another, by moving the stage on which the part sits, and a digital read out reports how far the stage moved to reach those points. Finally, the most technologically advanced methods involve software that analyzes the image and reports measurements. The first two methods are the most common; the third is newer and not as widespread, but its adoption is ongoing in the digital era.

The first commercial comparator was developed by James Hartness and Russell W. Porter. Hartness' long-continuing work as the Chairman of the U.S.'s National Screw-Thread Commission led him to apply his familiarity with optics (from his avocations of astronomy and telescope-building) to the problem of screw thread inspection. The Hartness Screw-Thread Comparator was for many years a profitable product for the Jones and Lamson Machine Company, of which he was president.

In subsequent decades optical comparators have been made by many companies and have been applied to the inspection of many kinds of parts. Today they may be found in many machine shops.

The idea of mixing optics and measurement and the use of the term comparator for metrological equipment, had existed in other forms prior to Hartness's work; but they had remained in realms of pure science (such as telescopy and microscopy) and highly specialized applied science (such as comparing master measuring standards). Hartness's comparator, intended for the routine inspection of machined parts, was a natural next step in the era during which applied science became widely integrated into industrial production.

Usage

Profile projector is widely used for complex shape stampings, gears, cams, threads and comparing the measured contour model. The profile projector is widely used in major machinery manufacturing including aviation, aerospace industry, watches and clocks, electronics, instrumentation industry, research institutes and detection metering stations at all levels and etc.

Work Principle

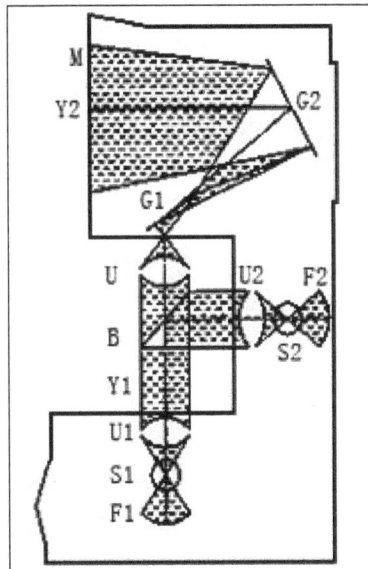

Profile projector working structure.

The projector magnifies the profile of the specimen, and displays this on the built-in projection screen. On this screen there is typically a grid that can be rotated 360 degrees so the X-Y axis of the screen can be aligned with a straight edge of the machined part to examine or measure. This projection screen displays the profile of the specimen and is magnified for better ease of calculating linear measurements. An edge of the specimen to examine may be lined up with the grid on the screen. From there, simple measurements may be taken for distances to other points. This is being done on a magnified profile of the specimen. It can be simpler as well as reduce errors by measuring on the magnified projection screen of a profile projector. The typical method for lighting is by diascopic illumination, which is lighting from behind. This type of lighting is also called transmitted illumination when the specimen is translucent and light can pass through it. If the specimen is opaque, then the light will not go through it, but will form a profile of the specimen. Measuring of the sample can be done on the projection screen. A profile projector may also have episcopic illumination (which is light shining from above). This useful in displaying bores or internal areas that may need to be measured.

Features

Projection Methods

- Vertical projectors: The main axis is parallel to the plane of the screen. They're most common, and suitable for flat parts or smaller work-pieces.

- Horizontal Projector: The main axis is perpendicular to the plane of the projection screen. They are mainly medium and large, and suitable for shaft parts or heavy work-pieces with large volume, although having a horizontal table below without a hole for light transmission can be convenient for small machines that use a silhouette lighting arrangement.

Positive or Inverted Images

For the simplest profile projector, the product's inverted image, also known as mirror image, will be displayed on the screen. In order to facilitate the measurement, sometimes we will deliberately add a plus-image system, changing the inverted image into a positive one, but it will undoubtedly increase the cost and reduced the measurement accuracy.

Screen Size

As for selection of screen size, you should carefully consider whether the entire part must be appeared on the screen. If it can achieve the purpose via segment observation, it's no need for large screen. Every projector manufacturer has a wide range of screen sizes to choose.

Magnification

The magnification of lens is fixed. Different part of measured piece often request different magnification. But the projector factory standard configuration is usually with only one lens, you must purchase according the real needs.

Work Table and Accessories

Work table is used to place and hold the measured piece. Its own volume, X, Y travel and carrying capacity are critical. Meanwhile, for the convenience of holding workpiece, it will be need to buy a rotary table, V-holder and other accessories.

Besides, the projector must also have a flexible and stable focusing mechanism and large working distance (the top surface of the workpiece and the lens pitch). Select appropriate data processing mode: all optical measuring projectors on market, without exception, have been digitized. We must focus on its data processing capabilities.

Precision

Current commercially available optical measuring projectors' theory accuracy, because the optical lens and gratings are on similar quality, are also similar. Therefore, there's no need to deliberately pursue high precision.

WATER JET CUTTER

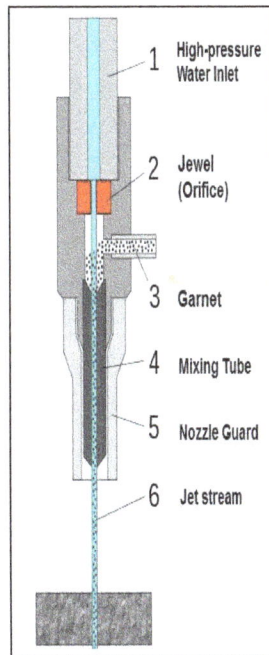

A diagram of a water jet cutter. 1) high-pressure water inlet. 2) jewel (ruby or diamond). 3) abrasive (garnet). 4) mixing tube. 5) guard. 6) cutting water jet. 7) cut material.

A water jet cutter, also known as a water jet or waterjet, is an industrial tool capable of cutting a wide variety of materials using a very high-pressure jet of water, or a mixture of water and an abrasive substance. The term abrasive jet refers specifically to the use of a mixture of water and abrasive to cut hard materials such as metal or granite, while the terms pure waterjet and water-only cutting refer to waterjet cutting without the use of added abrasives, often used for softer materials such as wood or rubber.

Waterjet cutting is often used during fabrication of machine parts. It is the preferred method when the materials being cut are sensitive to the high temperatures generated by other methods. Waterjet cutting is used in various industries, including mining and aerospace, for cutting, shaping, and reaming.

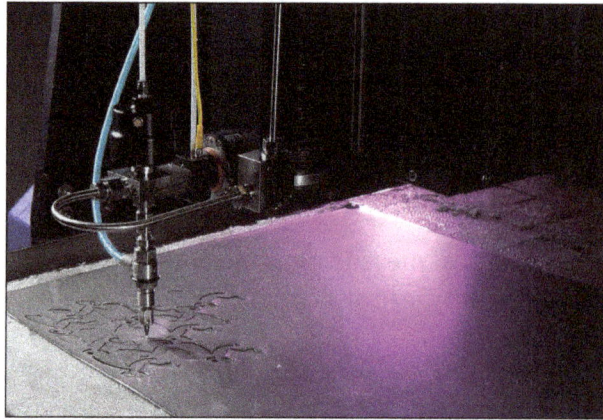

Waterjet CNC cutting machine.

Waterjet

While using high-pressure water for erosion dates back as far as the mid-1800s with hydraulic mining, it was not until the 1930s that narrow jets of water started to appear as an industrial cutting device. In 1933, the Paper Patents Company in Wisconsin developed a paper metering, cutting, and reeling machine that used a diagonally moving waterjet nozzle to cut a horizontally moving sheet of continuous paper. These early applications were at a low pressure and restricted to soft materials like paper.

Waterjet technology evolved in the post-war era as researchers around the world searched for new methods of efficient cutting systems. In 1956, Carl Johnson of Durox International in Luxembourg developed a method for cutting plastic shapes using a thin stream high-pressure waterjet, but those materials, like paper, were soft materials. In 1958, Billie Schwacha of North American Aviation developed a system using ultra-high-pressure liquid to cut hard materials. This system used a 100,000 psi (690 MPa) pump to deliver a hypersonic liquid jet that could cut high strength alloys such as PH15-7-MO stainless steel. Used to cut honeycomb laminate for the Mach 3 North American XB-70 Valkyrie, this cutting method resulted in delaminating at high speed, requiring changes to the manufacturing process.

While not effective for the XB-70 project, the concept was valid and further research continued to evolve waterjet cutting. In 1962, Philip Rice of Union Carbide explored using a pulsing waterjet at up to 50,000 psi (340 MPa) to cut metals, stone, and other materials. Research by S.J. Leach and G.L. Walker in the mid-1960s expanded on traditional coal waterjet cutting to determine ideal nozzle shape for high-pressure waterjet cutting of stone, and Norman Franz in the late 1960s focused on waterjet cutting of soft materials by dissolving long chain polymers in the water to improve the cohesiveness of the jet stream. In the early 1970s, the desire to improve the durability of the waterjet nozzle led Ray Chadwick, Michael Kurko, and Joseph Corriveau of the Bendix Corporation to come up with the idea of using corundum crystal to form a waterjet orifice, while Norman Franz expanded on this and created a waterjet nozzle with an orifice as small as 0.002 inches (0.051 mm) that operated at pressures up to 70,000 psi (480 MPa). John Olsen, along with George Hurlburt and Louis Kapcsandy at Flow Research (later Flow Industries), further improved the commercial potential of the waterjet by showing that treating the water beforehand could increase the operational life of the nozzle.

High Pressure

High-pressure vessels and pumps became affordable and reliable with the advent of steam power. By the mid-1800s, steam locomotives were common and the first efficient steam-driven fire engine was operational. By the turn of the century, high-pressure reliability improved, with locomotive research leading to a sixfold increase in boiler pressure, some reaching 1,600 psi (11 MPa). Most high-pressure pumps at this time, though, operated around 500–800 psi (3.4–5.5 MPa).

High-pressure systems were further shaped by the aviation, automotive, and oil industries. Aircraft manufacturers such as Boeing developed seals for hydraulically boosted control systems in the 1940s, while automotive designers followed similar research for hydraulic suspension systems. Higher pressures in hydraulic systems in the oil industry also led to the development of advanced seals and packing to prevent leaks.

These advances in seal technology, plus the rise of plastics in the post-war years, led to the development of the first reliable high-pressure pump. The invention of Marlex by Robert Banks and John Paul Hogan of the Phillips Petroleum company required a catalyst to be injected into the polyethylene. McCartney Manufacturing Company in Baxter Springs, Kansas, began manufacturing these high-pressure pumps in 1960 for the polyethylene industry. Flow Industries in Kent, Washington set the groundwork for commercial viability of waterjets with John Olsen's development of the high-pressure fluid intensifier in 1973, a design that was further refined in 1976. Flow Industries then combined the high-pressure pump research with their waterjet nozzle research and brought waterjet cutting into the manufacturing world.

Abrasive Waterjet

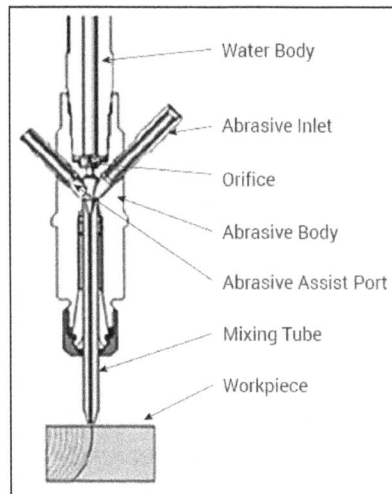

Abrasive Waterjet Nozzle.

While cutting with water is possible for soft materials, the addition of an abrasive turned the waterjet into a modern machining tool for all materials. This began in 1935 when the idea of adding an abrasive to the water stream was developed by Elmo Smith for the liquid abrasive blasting. Smith's design was further refined by Leslie Tirrell of the Hydroblast Corporation in 1937, resulting in a nozzle design that created a mix of high-pressure water and abrasive for the purpose of wet blasting.

The first publications on the modern Abrasive Waterjets (AWJ) cutting were published by Dr. Mohamed Hashish in the 1982 BHR proceedings showing, for the first time, that waterjets with relatively small amounts of abrasives are capable of cutting hard materials such as steel and concrete. Dr. Mohamed Hashish, was awarded a patent on forming AWJ in 1987. Dr. Hashish, who also coined the new term Abrasive Waterjet (AWJ), and his team continued to develop and improve the AWJ technology and its hardware for many applications which is now in over 50 industries worldwide. A most critical development was creating a durable mixing tube that could withstand the power of the high-pressure AWJ, and it was Boride Products (now Kennametal) development of their ROCTEC line of ceramic tungsten carbide composite tubes that significantly increased the operational life of the AWJ nozzle. Current work on AWJ nozzles is on micro abrasive waterjet so cutting with jets smaller than 0.015 inches (0.38 mm) in diameter can be commercialized.

Working with Ingersoll-Rand Waterjet Systems, Michael Dixon implemented the first production practical means of cutting titanium sheets an abrasive waterjet system very similar to those in widespread use today. By January 1989, that system was being run 24 hours a day producing titanium parts for the B-1B largely at Rockwell's North American Aviation facility in Newark, Ohio.

Waterjet Control

Large water jet abrasive cutting machine.

As waterjet cutting moved into traditional manufacturing shops, controlling the cutter reliably and accurately was essential. Early waterjet cutting systems adapted traditional systems such as mechanical pantographs and CNC systems based on John Parsons' 1952 NC milling machine and running G-code. Challenges inherent to waterjet technology revealed the inadequacies of traditional G-Code, as accuracy depends on varying the speed of the nozzle as it approaches corners and details. Creating motion control systems to incorporate those variables became a major innovation for leading waterjet manufacturers in the early 1990s, with Dr John Olsen of OMAX Corporation developing systems to precisely position the waterjet nozzle while accurately specifying the speed at every point along the path, and also utilizing common PCs as a controller. The largest waterjet manufacturer, Flow International (a spinoff of Flow Industries), recognized the benefits of that system and licensed the OMAX software, with the result that the vast majority of waterjet cutting machines worldwide are simple to use, fast, and accurate.

Operation

All waterjets follow the same principle of using high pressure water focused into a beam by a nozzle. Most machines accomplish this by first running the water through a high pressure pump. There are two types of pumps used to create this high pressure; an intensifier pump and a direct drive or crankshaft pump. A direct drive pump works much like a car engine, forcing water through high pressure tubing using plungers attached to a crankshaft. An intensifier pump creates pressure by using hydraulic oil to move a piston forcing the water through a tiny hole. The water then travels along the high pressure tubing to the nozzle of the waterjet. In the nozzle, the water is focused into a thin beam by a jewel orifice. This beam of water is ejected from the nozzle, cutting through the material by spraying it with the jet of speed on the order of Mach 3, around 2,500 ft/s (760 m/s). The process is the same for abrasive waterjets until the water reaches the nozzle. Here abrasives such as garnet and aluminium oxide, are fed into the nozzle via an abrasive inlet. The abrasive then mixes with the water in a mixing tube and is forced out the end at high pressure.

Benefits

An important benefit of the water jet is the ability to cut material without interfering with its inherent structure, as there is no heat-affected zone (HAZ). Minimizing the effects of heat allows metals to be cut without harming or changing intrinsic properties. Sharp corners, bevels, pierce holes, and shapes with minimal inner radii are all possible.

Water jet cutters are also capable of producing intricate cuts in material. With specialized software and 3-D machining heads, complex shapes can be produced.

The kerf, or width, of the cut can be adjusted by swapping parts in the nozzle, as well as changing the type and size of abrasive. Typical abrasive cuts have a kerf in the range of 0.04 to 0.05 in (1.0–1.3 mm), but can be as narrow as 0.02 inches (0.51 mm). Non-abrasive cuts are normally 0.007 to 0.013 in (0.18–0.33 mm), but can be as small as 0.003 inches (0.076 mm), which is approximately that of a human hair. These small jets can permit small details in a wide range of applications.

Water jets are capable of attaining accuracy down to 0.005 inches (0.13 mm) and repeatability down to 0.001 inches (0.025 mm).

Due to its relatively narrow kerf, water jet cutting can reduce the amount of scrap material produced, by allowing uncut parts to be nested more closely together than traditional cutting methods. Water jets use approximately 0.5 to 1 US gal (1.9–3.8 l) per minute (depending on the cutting head's orifice size), and the water can be recycled using a closed-loop system. Waste water usually is clean enough to filter and dispose of down a drain. The garnet abrasive is a non-toxic material that can be mostly recycled for repeated use; otherwise, it can usually be disposed in a landfill. Water jets also produce fewer airborne dust particles, smoke, fumes, and contaminants, reducing operator exposure to hazardous materials.

Meatcutting using waterjet technology eliminates the risk of cross contamination since there is no contact medium (namely, a blade).

Versatility

A water jet cutting a metal tool.

Because the nature of the cutting stream can be easily modified the water jet can be used in nearly every industry; there are many different materials that the water jet can cut. Some of them have unique characteristics that require special attention when cutting.

Materials commonly cut with a water jet include textiles, rubber, foam, plastics, leather, composites, stone, tile, glass, metals, food, paper and much more. Most ceramics can also be cut on an abrasive water jet as long as the material is softer than the abrasive being used (between 7.5 and 8.5 on the Mohs scale). Examples of materials that cannot be cut with a water jet are tempered glass and diamonds. Water jets are capable of cutting up to 6 in (150 mm) of metals and 18 in (460 mm) of most materials,, though in specialized coal mining applications, water jets are capable of cutting up to 100 ft (30 m) using a 1 in (25 mm) nozzle.

Specially designed water jet cutters are commonly used to remove excess bitumen from road surfaces that have become the subject of binder flushing. Flushing is a natural occurrence caused during hot weather where the aggregate becomes level with the bituminous binder layer creating a hazardously smooth road surface during wet weather.

Availability

Commercial water jet cutting systems are available from manufacturers all over the world, in a range of sizes, and with water pumps capable of a range of pressures. Typical water jet cutting machines have a working envelope as small as a few square feet, or up to hundreds of square feet. Ultra-high-pressure water pumps are available from as low as 40,000 psi (280 MPa) up to 100,000 psi (690 MPa).

Process

There are six main process characteristics to water jet cutting:

- Uses a high velocity stream of Ultra High Pressure Water 30,000–90,000 psi (210–620 MPa) which is produced by a high pressure pump with possible abrasive particles suspended in the stream.

- Is used for machining a large array of materials, including heat-sensitive, delicate or very hard materials.

- Produces no heat damage to workpiece surface or edges.

- Nozzles are typically made of sintered boride or composite tungsten carbide.

- Produces a taper of less than 1 degree on most cuts, which can be reduced or eliminated entirely by slowing down the cut process or tilting the jet.

- Distance of nozzle from workpiece affects the size of the kerf and the removal rate of material. Typical distance is .125 in (3.2 mm).

Temperature is not as much of a factor.

Edge Quality

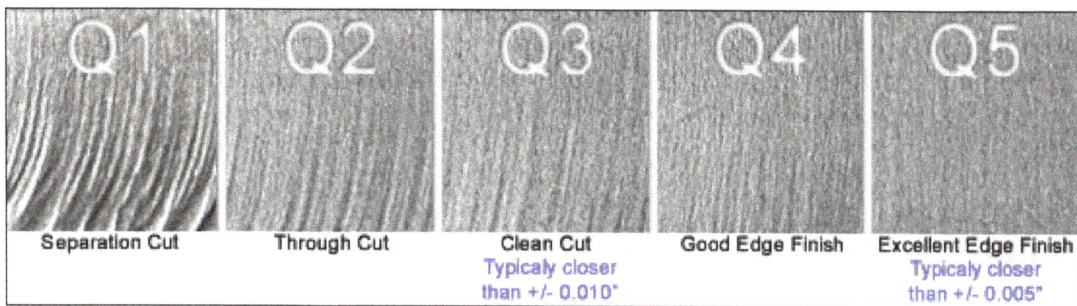

Q1	Q2	Q3	Q4	Q5
Separation Cut	Through Cut	Clean Cut Typically closer than +/- 0.010"	Good Edge Finish	Excellent Edge Finish Typically closer than +/- 0.005"

Edge quality for water jet cut parts is defined with the quality numbers Q1 through Q5. Lower numbers indicate rougher edge finish; higher numbers are smoother. For thin materials, the difference in cutting speed for Q1 could be as much as 3 times faster than the speed for Q5. For thicker materials, Q1 could be 6 times faster than Q5. For example, 4 inches (100 mm) thick aluminium Q5 would be 0.72 in/min (18 mm/min) and Q1 would be 4.2 in/min (110 mm/min), 5.8 times faster.

Multi-axis Cutting

A 5-Axis Waterjet Cutting Head.

A 5-Axis Waterjet Part.

In 1987, Ingersoll-Rand Waterjet Systems offered a 5-axis pure-water waterjet cutting system called the Robotic Waterjet System. The system was an overhead gantry design, similar in overall size to the HS-1000.

With recent advances in control and motion technology, 5-axis water jet cutting (abrasive and pure) has become a reality. Where the normal axes on a water jet are named Y (back/forth), X (left/right) and Z (up/down), a 5-axis system will typically add an A axis (angle from perpendicular) and C axis (rotation around the Z-axis). Depending on the cutting head, the maximum cutting angle for the A axis can be anywhere from 55, 60, or in some cases even 90 degrees from vertical. As such, 5-axis cutting opens up a wide range of applications that can be machined on a water jet cutting machine.

A 5-axis cutting head can be used to cut 4-axis parts, where the bottom surface geometries are shifted a certain amount to produce the appropriate angle and the Z-axis remains at one height. This can be useful for applications like weld preparation where a bevel angle needs to be cut on all sides of a part that will later be welded, or for taper compensation purposes where the kerf angle is transferred to the waste material thus eliminating the taper commonly found on water jet-cut parts. A 5-axis head can cut parts where the Z-axis is also moving along with all the other axes. This full 5-axis cutting could be used for cutting contours on various surfaces of formed parts.

Because of the angles that can be cut, part programs may need to have additional cuts to free the part from the sheet. Attempting to slide a complex part at a severe angle from a plate can be difficult without appropriate relief cuts.

LATHE

A watchmaker using a lathe to prepare a component cut from copper for a watch.

A lathe is a machine tool that rotates a workpiece about an axis of rotation to perform various operations such as cutting, sanding, knurling, drilling, deformation, facing, and turning, with tools that are applied to the workpiece to create an object with symmetry about that axis.

Uses

Lathes are used in woodturning, metalworking, metal spinning, thermal spraying, parts reclamation, and glass-working. Lathes can be used to shape pottery, the best-known design being the Potter's wheel. Most suitably equipped metalworking lathes can also be used to produce most solids of revolution, plane surfaces and screw threads or helices. Ornamental lathes can produce three-dimensional solids of incredible complexity. The workpiece is usually held in place by either one or two centers, at least one of which can typically be moved horizontally to accommodate varying workpiece lengths. Other work-holding methods include clamping the work about the axis of rotation using a chuck or collet, or to a faceplate, using clamps or dog clutch.

Products made by Lathes

Examples of objects that can be produced on a lathe include screws, candlesticks, gun barrels, cue sticks, table legs, bowls, baseball bats, musical instruments (especially woodwind instruments), crankshafts and much more.

Parts

A lathe may or may not have legs also known as a nugget, which sit on the floor and elevate the lathe bed to a working height. A lathe may be small and sit on a workbench or table, not requiring a stand.

Almost all lathes have a bed, which is (almost always) a horizontal beam (although CNC lathes commonly have an inclined or vertical beam for a bed to ensure that swarf, or chips, falls free of the bed). Woodturning lathes specialized for turning large bowls often have no bed or tail stock, merely a free-standing headstock and a cantilevered tool rest.

At one end of the bed (almost always the left, as the operator faces the lathe) is a headstock. The headstock contains high-precision spinning bearings. Rotating within the bearings is a horizontal axle, with an axis parallel to the bed, called the spindle. Spindles are often hollow and have exterior threads and/or an interior Morse taper on the "inboard" (i.e., facing to the right / towards the bed) by which work-holding accessories may be mounted to the spindle. Spindles may also have exterior threads and/or an interior taper at their "outboard" (i.e., facing away from the bed) end, and/or may have a hand-wheel or other accessory mechanism on their outboard end. Spindles are powered and impart motion to the workpiece.

The spindle is driven either by foot power from a treadle and flywheel or by a belt or gear drive to a power source. In most modern lathes this power source is an integral electric motor, often either in the headstock, to the left of the headstock, or beneath the headstock, concealed in the stand.

In addition to the spindle and its bearings, the headstock often contains parts to convert the motor speed into various spindle speeds. Various types of speed-changing mechanism achieve this, from a cone pulley or step pulley, to a cone pulley with back gear (which is essentially a low range, similar in net effect to the two-speed rear of a truck), to an entire gear train similar to that of a manual-shift auto transmission. Some motors have electronic rheostat-type speed controls, which obviates cone pulleys or gears.

The counterpoint to the headstock is the tailstock, sometimes referred to as the loose head, as it can be positioned at any convenient point on the bed by sliding it to the required area. The tailstock contains a barrel, which does not rotate, but can slide in and out parallel to the axis of the bed and directly in line with the headstock spindle. The barrel is hollow and usually contains a taper to facilitate the gripping of various types of tooling. Its most common uses are to hold a hardened steel center, which is used to support long thin shafts while turning, or to hold drill bits for drilling axial holes in the work piece. Many other uses are possible.

Metalworking lathes have a carriage (comprising a saddle and apron) topped with a cross-slide, which is a flat piece that sits crosswise on the bed and can be cranked at right angles to the bed. Sitting atop the cross slide is usually another slide called a compound rest, which provides 2 additional axes of motion, rotary and linear. Atop that sits a toolpost, which holds a cutting tool, which removes material from the workpiece. There may or may not be a leadscrew, which moves the cross-slide along the bed.

Woodturning and metal spinning lathes do not have cross-slides, but rather have banjos, which are flat pieces that sit crosswise on the bed. The position of a banjo can be adjusted by hand; no gearing is involved. Ascending vertically from the banjo is a tool-post, at the top of which is a horizontal tool-rest. In woodturning, hand tools are braced against the tool rest and levered into the workpiece. In metal spinning, the further pin ascends vertically from the tool rest and serves as a fulcrum against which tools may be levered into the workpiece.

Accessories

A steady rest.

Unless a workpiece has a taper machined onto it which perfectly matches the internal taper in the spindle, or has threads which perfectly match the external threads on the spindle (two conditions which rarely exist), an accessory must be used to mount a workpiece to the spindle.

A workpiece may be bolted or screwed to a faceplate, a large, flat disk that mounts to the spindle. In the alternative, faceplate dogs may be used to secure the work to the faceplate.

A workpiece may be mounted on a mandrel, or circular work clamped in a three- or four-jaw chuck. For irregular shaped workpieces it is usual to use a four jaw (independent moving jaws) chuck. These holding devices mount directly to the lathe headstock spindle.

In precision work, and in some classes of repetition work, cylindrical workpieces are usually held in a collet inserted into the spindle and secured either by a draw-bar, or by a collet closing cap on the spindle. Suitable collets may also be used to mount square or hexagonal workpieces. In precision toolmaking work such collets are usually of the draw-in variety, where, as the collet is tightened, the workpiece moves slightly back into the headstock, whereas for most repetition work the dead length variety is preferred, as this ensures that the position of the workpiece does not move as the collet is tightened.

A soft workpiece (e.g., wood) may be pinched between centers by using a spur drive at the headstock, which bites into the wood and imparts torque to it.

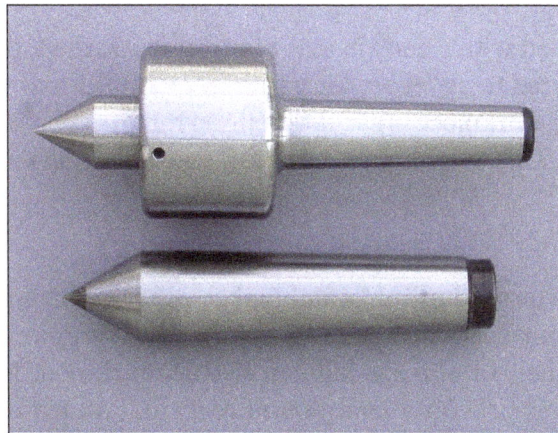

Live center (top); dead center (bottom).

A soft dead center is used in the headstock spindle as the work rotates with the centre. Because the centre is soft it can be trued in place before use. The included angle is 60°. Traditionally, a hard dead center is used together with suitable lubricant in the tailstock to support the workpiece. In modern practice the dead center is frequently replaced by a live center, as it turns freely with the workpiece usually on ball bearings reducing the frictional heat, especially important at high speeds. When clear facing a long length of material it must be supported at both ends. This can be achieved by the use of a traveling or fixed steady. If a steady is not available, the end face being worked on may be supported by a dead (stationary) half center. A half center has a flat surface machined across a broad section of half of its diameter at the pointed end. A small section of the tip of the dead center is retained to ensure concentricity. Lubrication must be applied at this point of contact and tail stock pressure reduced. A lathe carrier or lathe dog may also be employed when turning between two centers.

In woodturning, one variation of a live center is a cup center, which is a cone of metal surrounded by an annular ring of metal that decreases the chances of the workpiece splitting.

A circular metal plate with even spaced holes around the periphery, mounted to the spindle, is called an "index plate". It can be used to rotate the spindle to a precise angle, then lock it in place, facilitating repeated auxiliary operations done to the workpiece.

Other accessories, including items such as taper turning attachments, knurling tools, vertical slides, fixed and traveling steadies, etc., increase the versatility of a lathe and the range of work it may perform.

Modes of Use

When a workpiece is fixed between the headstock and the tail-stock, it is said to be "between centers". When a workpiece is supported at both ends, it is more stable, and more force may be applied to the workpiece, via tools, at a right angle to the axis of rotation, without fear that the workpiece may break loose.

When a workpiece is fixed only to the spindle at the headstock end, the work is said to be "face work". When a workpiece is supported in this manner, less force may be applied to the workpiece, via tools, at a right angle to the axis of rotation, lest the workpiece rip free. Thus, most work must be done axially, towards the headstock, or at right angles, but gently.

When a workpiece is mounted with a certain axis of rotation, worked, then remounted with a new axis of rotation, this is referred to as "eccentric turning" or "multi-axis turning". The result is that various cross sections of the workpiece are rotationally symmetric, but the workpiece as a whole is not rotationally symmetric. This technique is used for camshafts, various types of chair legs.

Varieties

The smallest lathes are "jewelers lathes" or "watchmaker lathes", which, though often small enough to be held in one hand are normally fastened to a bench. The workpieces machined on a jeweler's lathe are often metal, but other softer materials can also be machined. Jeweler's lathes can be used with hand-held "graver" tools or with a "compound rest" that attach to the lathe bed and allows the tool to be clamped in place and moved by a screw or lever feed. Graver tools are generally supported by a T-rest, not fixed to a cross slide or compound rest. The work is usually held in a collet, but high-precision 3 and 6-jaw chucks are also commonly employed. Common spindle bore sizes are 6 mm, 8 mm and 10 mm. The term WW refers to the Webster/Whitcomb collet and lathe, invented by the American Watch Tool Company of Waltham, Massachusetts. Most lathes commonly referred to as watchmakers lathes are of this design. In 1909, the American Watch Tool company introduced the Magnus type collet (a 10-mm body size collet) using a lathe of the same basic design, the Webster/Whitcomb Magnus. (F.W.Derbyshire, Inc. retains the trade names Webster/Whitcomb and Magnus and still produces these collets.) Two bed patterns are common: the WW (Webster Whitcomb) bed, a truncated triangular prism (found only on 8 and 10 mm watchmakers' lathes); and the continental D-style bar bed (used on both 6 mm and 8 mm lathes by firms such as Lorch and Star). Other bed designs have been used, such a triangular prism on some Boley 6.5 mm lathes, and a V-edged bed on IME's 8 mm lathes.

Smaller metalworking lathes that are larger than jewelers' lathes and can sit on a bench or table, but offer such features as tool holders and a screw-cutting gear train are called hobby lathes, and larger versions, "bench lathes" this term also commonly applied to a special type of high-precision lathe used by toolmakers for one-off jobs. Even larger lathes offering similar features for producing or modifying individual parts are called "engine lathes". Lathes of these types do not have additional integral features for repetitive production, but rather are used for individual part production or modification as the primary role.

Lathes of this size that are designed for mass manufacture, but not offering the versatile screw-cutting capabilities of the engine or bench lathe, are referred to as "second operation" lathes.

Lathes with a very large spindle bore and a chuck on both ends of the spindle are called "oil field lathes".

Fully automatic mechanical lathes, employing cams and gear trains for controlled movement, are called screw machines.

Lathes that are controlled by a computer are CNC lathes.

Lathes with the spindle mounted in a vertical configuration, instead of horizontal configuration, are called vertical lathes or vertical boring machines. They are used where very large diameters must be turned, and the workpiece (comparatively) is not very long.

A lathe with a tool post that can rotate around a vertical axis, so as to present different tools towards the headstock (and the workpiece) are turret lathes.

A lathe equipped with indexing plates, profile cutters, spiral or helical guides, etc., so as to enable ornamental turning is an ornamental lathe.

Various combinations are possible: for example, a vertical lathe can have CNC capabilities as well (such as a CNC VTL).

Lathes can be combined with other machine tools, such as a drill press or vertical milling machine. These are usually referred to as combination lathes.

Major Categories

Woodworking Lathes

A modern woodworking lathe.

Woodworking lathes are the oldest variety. All other varieties are descended from these simple lathes. An adjustable horizontal metal rail the tool rest between the material and the operator accommodates the positioning of shaping tools, which are usually hand-held. After shaping, it is common practice to press and slide sandpaper against the still-spinning object to smooth the surface made with the metal shaping tools. The tool rest is usually removed during sanding, as it may be unsafe to have the operators hands between it and the spinning wood.

Many woodworking lathes can also be used for making bowls and plates. The bowl or plate needs only to be held at the bottom by one side of the lathe. It is usually attached to a metal face plate attached to the spindle. With many lathes, this operation happens on the left side of the headstock, where are no rails and therefore more clearance. In this configuration, the piece can be shaped inside and out. A specific curved tool rest may be used to support tools while shaping the inside.

Most woodworking lathes are designed to be operated at a speed of between 200 and 1,400 revolutions per minute, with slightly over 1,000 rpm considered optimal for most such work, and with larger workpieces requiring lower speeds.

Duplicating Lathes

Water-powered Blanchard lathe used for duplicating gun stocks.

One type of specialized lathe is duplicating or copying lathe also known as Blanchard lathe after its inventor Thomas Blanchard. This type of lathe was able to create shapes identical to a standard pattern and it revolutionized the process of gun stock making in 1820's when it was invented.

Patternmaker's Lathes

Patternmaker's double lathe.

Used to make a pattern for foundries, often from wood, but also plastics. A patternmaker's lathe looks like a heavy wood lathe, often with a turret and either a leadscrew or a rack and pinion to manually position the turret. The turret is used to accurately cut straight lines. They often have a provision to turn very large parts on the other end of the headstock, using a free-standing toolrest. Another way of turning large parts is a sliding bed, which can slide away from the headstock and thus open up a gap in front of the headstock for large parts.

Metalworking Lathes

In a metalworking lathe, metal is removed from the workpiece using a hardened cutting tool, which is usually fixed to a solid moveable mounting, either a tool-post or a turret, which is then moved against the workpiece using handwheels and/or computer-controlled motors. These cutting tools come in a wide range of sizes and shapes, depending upon their application. Some common styles are diamond, round, square and triangular.

A CNC metalworking lathe.

The tool-post is operated by lead-screws that can accurately position the tool in a variety of planes. The tool-post may be driven manually or automatically to produce the roughing and finishing cuts required to turn the workpiece to the desired shape and dimensions, or for cutting threads, worm gears, etc. Cutting fluid may also be pumped to the cutting site to provide cooling, lubrication and clearing of swarf from the workpiece. Some lathes may be operated under control of a computer for mass production of parts.

Manually controlled metalworking lathes are commonly provided with a variable-ratio gear-train to drive the main lead-screw. This enables different thread pitches to be cut. On some older lathes or more affordable new lathes, the gear trains are changed by swapping gears with various numbers of teeth onto or off of the shafts, while more modern or expensive manually controlled lathes have a quick-change box to provide commonly used ratios by the operation of a lever. CNC lathes use computers and servomechanisms to regulate the rates of movement.

On manually controlled lathes, the thread pitches that can be cut are, in some ways, determined by the pitch of the lead-screw: A lathe with a metric lead-screw will readily cut metric threads (including BA), while one with an imperial lead-screw will readily cut imperial-unit-based threads such as BSW or UTS (UNF, UNC). This limitation is not insurmountable, because a 127-tooth gear, called a transposing gear, is used to translate between metric and inch thread pitches. However, this is optional equipment that many lathe owners do not own. It is also a larger change-wheel than the others, and on some lathes may be larger than the change-wheel mounting banjo is capable of mounting.

The workpiece may be supported between a pair of points called centres, or it may be bolted to a faceplate or held in a chuck. A chuck has movable jaws that can grip the workpiece securely.

There are some effects on material properties when using a metalworking lathe. There are few chemical or physical effects, but there are many mechanical effects, which include residual stress, micro-cracks, work-hardening, and tempering in hardened materials.

Cue Lathes

Cue lathes function similarly to turning and spinning lathes, allowing a perfectly radially-symmetrical cut for billiard cues. They can also be used to refinish cues that have been worn over the years.

Glass-working Lathes

Glass-working lathes are similar in design to other lathes, but differ markedly in how the workpiece is modified. Glass-working lathes slowly rotate a hollow glass vessel over a fixed- or variable-temperature flame. The source of the flame may be either hand-held or mounted to a banjo/cross-slide that can be moved along the lathe bed. The flame serves to soften the glass being worked, so that the glass in a specific area of the workpiece becomes ductile and subject to forming either by inflation ("glassblowing") or by deformation with a heat-resistant tool. Such lathes usually have two head-stocks with chucks holding the work, arranged so that they both rotate together in unison. Air can be introduced through the headstock chuck spindle for glassblowing. The tools to deform the glass and tubes to blow (inflate) the glass are usually handheld.

In diamond turning, a computer-controlled lathe with a diamond-tipped tool is used to make precision optical surfaces in glass or other optical materials. Unlike conventional optical grinding, complex aspheric surfaces can be machined easily. Instead of the dovetailed ways used on the tool slide of a metal-turning lathe, the ways typically float on air bearings, and the position of the tool is measured by optical interferometry to achieve the necessary standard of precision for optical work. The finished work piece usually requires a small amount of subsequent polishing by conventional techniques to achieve a finished surface suitably smooth for use in a lens, but the rough grinding time is significantly reduced for complex lenses.

Metal-spinning Lathes

In metal spinning, a disk of sheet metal is held perpendicularly to the main axis of the lathe, and tools with polished tips (spoons) or roller tips are hand-held, but levered by hand against fixed posts, to develop pressure that deforms the spinning sheet of metal.

Metal-spinning lathes are almost as simple as wood-turning lathes. Typically, metal spinning requires a mandrel, usually made from wood, which serves as the template onto which the workpiece is formed (asymmetric shapes can be made, but it is a very advanced technique). For example, to make a sheet metal bowl, a solid block of wood in the shape of the bowl is required; similarly, to make a vase, a solid template of the vase is required.

Given the advent of high-speed, high-pressure, industrial die forming, metal spinning is less common now than it once was, but still a valuable technique for producing one-off prototypes or small batches, where die forming would be uneconomical.

Ornamental Turning Lathes

The ornamental turning lathe was developed around the same time as the industrial screw-cutting lathe in the nineteenth century. It was used not for making practical objects, but for decorative work – ornamental turning. By using accessories such as the horizontal and vertical cutting frames, eccentric chuck and elliptical chuck, solids of extraordinary complexity may be produced by various generative procedures.

A special-purpose lathe, the Rose engine lathe, is also used for ornamental turning, in particular for engine turning, typically in precious metals, for example to decorate pocket-watch cases. As well as a wide range of accessories, these lathes usually have complex dividing arrangements to allow the exact rotation of the mandrel. Cutting is usually carried out by rotating cutters, rather than directly by the rotation of the work itself. Because of the difficulty of polishing such work, the materials turned, such as wood or ivory, are usually quite soft, and the cutter has to be exceptionally sharp. The finest ornamental lathes are generally considered to be those made by Holtzapffel around the turn of the 19th century.

Reducing Lathe

Many types of lathes can be equipped with accessory components to allow them to reproduce an item: the original item is mounted on one spindle, the blank is mounted on another, and as both turn in synchronized manner, one end of an arm "reads" the original and the other end of the arm "carves" the duplicate.

A reduction lathe is a specialized lathe that is designed with this feature and incorporates a mechanism similar to a pantograph, so that when the "reading" end of the arm reads a detail that measures one inch (for example), the cutting end of the arm creates an analogous detail that is (for example) one quarter of an inch (a 4:1 reduction, although given appropriate machinery and appropriate settings, any reduction ratio is possible).

Reducing lathes are used in coin-making, where a plaster original (or an epoxy master made from the plaster original, or a copper-shelled master made from the plaster original, etc.) is duplicated and reduced on the reducing lathe, generating a master die.

Rotary Lathes

A lathe in which softwood, like spruce or pine, or hardwood, like birch, logs are turned against a very sharp blade and peeled off in one continuous or semi-continuous roll. Invented by Immanuel Nobel (father of the more famous Alfred Nobel). The first such lathes in the United States were set up in the mid-19th century. The product is called wood veneer and it is used for making plywood and as a cosmetic surface veneer on some grades of chipboard.

Watchmaker's Lathes

Watchmakers lathes are delicate but precise metalworking lathes, usually without provision for screwcutting, and are still used by horologists for work such as the turning of balance staffs. A handheld tool called a graver is often used in preference to a slide-mounted tool. The original watchmaker's turns was a simple dead-center lathe with a moveable rest and two loose head-stocks. The workpiece would be rotated by a bow, typically of horsehair, wrapped around it.

Transcription or Recording, Lathes

Transcription or recording lathes are used to make grooves on a surface for recording sounds. These were used in creating sound grooves on wax cylinders and then on flat recording discs originally also made of wax, but later as lacquers on a substrata. Originally the cutting lathes were driven by sound vibrations through a horn in a process known as Acoustic recording and later driven by an electric current when microphones were first used in sound recording. Many such lathes were professional models, but others were developed for home recording and were common before the advent of home tape recording.

Gallery

Examples of lathes

Large old lathe.

Small metalworking lathe.

Turned chess pieces.

Performance Evaluation

National and international standards are used to standardize the definitions, environmental requirements, and test methods used for the performance evaluation of lathes. Election of the standard to be used is an agreement between the supplier and the user and has some significance in the design of the lathe. In the United States, ASME has developed the B5.57 Standard entitled "Methods for Performance Evaluation of Computer Numerically Controlled Lathes and Turning Centers", which establishes requirements and methods for specifying and testing the performance of CNC lathes and turning centers.

PATTERN

Wooden pattern for a cast-iron gear with curved spokes.

The top and bottom halves of a sand casting mould showing the cavity prepared by patterns.
Cores to accommodate holes can be seen in the bottom half of the mould,
which is called the drag. The top half of the mould is called the cope.

In casting, a pattern is a replica of the object to be cast, used to prepare the cavity into which molten material will be poured during the casting process.

Patterns used in sand casting may be made of wood, metal, plastics or other materials. Patterns are made to exacting standards of construction, so that they can last for a reasonable length of time, according to the quality grade of the pattern being built, and so that they will repeatably provide a dimensionally acceptable casting.

Patternmaking

The making of patterns, called patternmaking (sometimes styled pattern-making or pattern making), is a skilled trade that is related to the trades of tool and die making and moldmaking, but also often incorporates elements of fine woodworking. Patternmakers (sometimes styled pattern-makers or pattern makers) learn their skills through apprenticeships and trade schools over many years of experience. Although an engineer may help to design the pattern, it is usually a patternmaker who executes the design.

Materials Used

Typically, materials used for pattern making are wood, metal or plastics. Wax and Plaster of Paris are also used, but only for specialized applications. Sugar pine is the most commonly used material for patterns, primarily because it is soft, light, and easy to work. Honduras Mahogany was used for more production parts because it is harder and would last longer than pine. Once properly cured it is about as stable as any wood available, not subject to warping or curling. Once the pattern is built the foundry does not want it changing shape. True Honduras Mahogany is harder to find now because of the decimation of the rain forests, so now there are a variety of woods marketed as Mahogany. Fiberglass and plastic patterns have gained popularity in recent years because they are water proof and very durable. Metal patterns are long lasting and do not succumb to moisture, but they are heavier, more expensive and difficult to repair once damaged.

Wax patterns are used in a casting process called investment casting. A combination of paraffin wax, bees wax and carnauba wax is used for this purpose.

Plaster of paris is usually used in making master dies and molds, as it gains hardness quickly, with a lot of flexibility when in the setting stage.

Design

Sprues, Gates, Risers, Cores and Chills

The patternmaker or foundry engineer decides where the sprues, gating systems, and risers are placed with respect to the pattern. Where a hole is desired in a casting, a core may be used which defines a volume or location in a casting where metal will not flow into. Sometimes chills may be placed on a pattern surface prior to molding, which are then formed into the sand mould. Chills are heat sinks which enable localized rapid cooling. The rapid cooling may be desired to refine the grain structure or determine the freezing sequence of the molten metal which is poured into the mould. Because they are at a much cooler temperature, and often a different metal than what is being poured, they do not attach to the casting when the casting cools. The chills can then be reclaimed and reused.

The design of the feeding and gating system is usually referred to as methoding or methods design. It can be carried out manually, or interactively using general-purpose CAD software, or semi-automatically using special-purpose software (such as AutoCAST)

Types of Patterns

Patterns are made of wood, metal, ceramic, or hard plastics and vary in complexity.

A single piece pattern, or loose pattern, is the simplest. It is a replica of the desired casting—usually in a slightly larger size to offset the shrinkage of the intended metal. Gated patterns connect a number of loose patterns together with a series of runners that will be detached after shake-out. Segmented or multi-piece patterns create a casting in several pieces to be joined in post-processing.

Match plate patterns are patterns with the top and bottom parts of the pattern, also known as the cope and drag portions, mounted on opposite sides of a board. This adaptation allows patterns to be quickly pressed into the molding material. A similar technique called a cope and drag pattern is often used for large castings or large production runs: in this variation, the two sides of the pattern are mounted on separate pattern plates that can be hooked up to horizontal or vertical machines and pressed into the molding material. When the parting lines between the cope and drag are irregular, a follow board can be used to support irregularly shaped, loose patterns.

Sweep patterns are used for symmetric molds, which are contoured shapes rotated around a center axis or pole through the molding material. A sweep pattern is a form of skeleton pattern: any geometrical pattern that creates a mold by being moved through the molding material.

Allowances

To compensate for any dimensional and structural changes which will happen during the casting or patterning process, allowances are usually made in the pattern.

Contraction Allowance/Shrinkage Allowance

Shrinkage Allowance: Almost all metals shrink or contract volumetrically after solidification to obtain a particular size of casting an amount is equal to the shrinkage or contraction.

The metal will undergo shrinkage during solidification and contract further on cooling to room temperature. To compensate this, the pattern is made larger than the required casting. This extra size is given on the pattern for metal shrinkage is called shrinkage allowance.

The pattern needs to incorporate suitable allowances for shrinkage; these are called contraction allowances, and their exact values depend on the alloy being cast and the exact sand casting method being used. Some alloys will have overall linear shrinkage of up to 2.5%, whereas other alloys may actually experience no shrinkage or a slight "positive" shrinkage or increase in size in the casting process (type metal and certain cast irons). The shrinkage amount is also dependent on the sand casting process employed, for example clay-bonded sand, chemical bonded sands, or other bonding materials used within the sand. This was traditionally accounted for using a shrink rule, which is an oversized rule.

Shrinkage can again be classified into liquid shrinkage and solid shrinkage. Liquid shrinkage is the reduction in volume during the process of solidification, and Solid shrinkage is the reduction in volume during the cooling of the cast metal. Shrinkage allowance takes into account only the solid shrinkage. The liquid shrinkage is accounted for by risers.

Draft Allowance

When the pattern is to be removed from the sand mold, there is a possibility that any leading edges may break off, or get damaged in the process. To avoid this, a taper is provided on the pattern, so as to facilitate easy removal of the pattern from the mold, and hence reduce damage to edges. The taper angle provided is called the Draft angle. The value of the draft angle depends upon the complexity of the pattern, the type of molding (hand molding or machine molding), height of the surface, etc. Draft provided on the casting is usually 1 to 3 degrees on external surfaces (5 to 8 internal surfaces).

Finishing or Machining Allowance

The surface finish obtained in sand castings is generally poor (dimensionally inaccurate), and hence in many cases, the cast product is subjected to machining processes like turning or grinding in order to improve the surface finish. During machining processes, some metal is removed from the piece. To compensate for this, a machining allowance (additional material) should be given in the casting. the amount of finish allowance depends on the material of the casting,size of casting,volume of production, method of molding, and etc..

Shake Allowance

Usually during removal of the pattern from the mold cavity, the pattern is rapped all around the faces, in order to facilitate easy removal. In this process, the final cavity is enlarged. To compensate for this, the pattern dimensions need to be reduced. There are no standard values for this allowance, as it is heavily dependent on the personnel. This allowance is a negative allowance, and a common way of going around this allowance is to increase the draft allowance. Shaking of the pattern causes an enlargement of the mould cavity and results in a bigger casting.

Distortion Allowance

During cooling of the mould, stresses developed in the solid metal may induce distortions in the cast. This is more evident when the mould is thinner in width as compared to its length. This can be eliminated by initially distorting the pattern in the opposite direction.

Demand

Patterns continue to be needed for sand casting of metals. For the production of gray iron, ductile iron and steel castings, sand casting remains the most widely used process. For aluminum castings, sand casting represents about 12% of the total tonnage by weight (surpassed only by die casting at 57%, and semi-permanent and permanent mold at 19%; based on 2006 shipments). The exact process and pattern equipment is always determined by the order quantities and the casting design. Sand casting can produce as little as one part, or as many as a million copies.

Although additive manufacturing modalities such as SLS or SLM have potential to replace casting for some production situations, casting is still far from being completely displaced. Wherever it provides suitable material properties at competitive unit cost, it will remain in demand.

Jig

A bicycle frame building jig.

A jig is a type of custom-made tool used to control the location and/or motion of parts or other tools.

Device with grooves and chucks.

A jig's primary purpose is to provide repeatability, accuracy, and interchangeability in the manufacturing of products. A jig is often confused with a fixture; a fixture holds the work in a fixed location. A device that does both functions (holding the work and guiding a tool) is called a jig.

An example of a jig is when a key is duplicated; the original is used as a jig so the new key can have the same path as the old one. Since the advent of automation and computer numerical controlled (CNC) machines, jigs are often not required because the tool path is digitally programmed and stored in memory. Jigs may be made for reforming plastics.

Jigs or templates have been known long before the industrial age. There are many types of jigs, and each one is custom-tailored to do a specific job.

Drill Jig

A drill jig is a type of jig that expedites repetitive hole center location on multiple interchangeable parts by acting as a template to guide the twist drill or other boring device into the precise location of each intended hole center. In metalworking practice, typically a hardened drill bushing lines each hole on the jig plate to keep the tool from damaging the jig.

Drill jigs started falling into disuse with the invention of the jig borer. Since the widespread penetration of the manufacturing industry by CNC machine tools, in which servo controls are capable of moving the tool to the correct location automatically, the need for drill jigs (and for the jobs of the drill press operators who used them) is much less than it used to be.

Jewelry Jig

A jig used in making jewelry, a specific type of jig, is a plate or open frame for holding work and helping to shape jewelry components made out of wire or small sheets of metal. A jig in the jewelry making application is used to help establish a pattern for use in shaping the wire or sheets of metal. In the jewelry application, the shaping of the metal is done by hand or with simple hand tools like a hammer.

FIXTURE

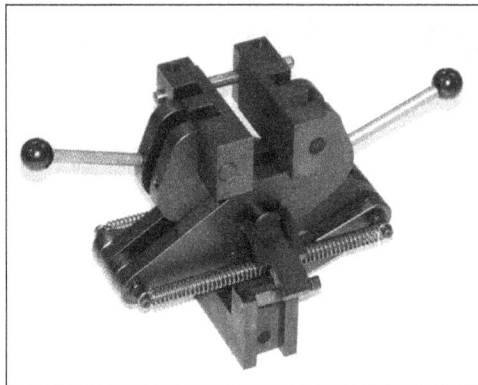

A common type of fixture, used in materials tensile testing.

A fixture is a work-holding or support device used in the manufacturing industry. Fixtures are used to securely locate (position in a specific location or orientation) and support the work, ensuring that all parts produced using the fixture will maintain conformity and interchangeability. Using a fixture improves the economy of production by allowing smooth operation and quick transition from part to part, reducing the requirement for skilled labor by simplifying how workpieces are mounted, and increasing conformity across a production run.

A fixture differs from a jig in that when a fixture is used, the tool must move relative to the workpiece; a jig moves the piece while the tool remains stationary.

Purpose

A fixture's primary purpose is to create a secure mounting point for a workpiece, allowing for support during operation and increased accuracy, precision, reliability, and interchangeability in the finished parts. It also serves to reduce working time by allowing quick set-up, and by smoothing the transition from part to part. It frequently reduces the complexity of a process, allowing for unskilled workers to perform it and effectively transferring the skill of the tool maker to the unskilled worker. Fixtures also allow for a higher degree of operator safety by reducing the concentration and effort required to hold a piece steady.

Economically speaking the most valuable function of a fixture is to reduce labor costs. Without a fixture, operating a machine or process may require two or more operators; using a fixture can eliminate one of the operators by securing the workpiece.

Design

These modular fixture components may be built into various arrangements to accommodate different workpieces.

Fixtures should be designed with economics in mind; the purpose of these devices is often to reduce costs, and so they should be designed in such a way that the cost reduction outweighs the cost of implementing the fixture. It is usually better, from an economic standpoint, for a fixture to result in a small cost reduction for a process in constant use, than for a large cost reduction for a process used only occasionally.

Most fixtures have a solid component, affixed to the floor or to the body of the machine and considered immovable relative to the motion of the machining bit, and one or more movable components known as clamps. These clamps (which may be operated by many different mechanical means) allow workpieces to be easily placed in the machine or removed, and yet stay secure during operation. Many are also adjustable, allowing for workpieces of different sizes to be used for different operations. Fixtures must be designed such that the pressure or motion of the machining operation (usually known as the feed) is directed primarily against the solid component of the fixture. This reduces the likelihood that the fixture will fail, interrupting the operation and potentially causing damage to infrastructure, components, or operators.

Fixtures may also be designed for very general or simple uses. These multi-use fixtures tend to be very simple themselves, often relying on the precision and ingenuity of the operator, as well as surfaces and components already present in the workshop, to provide the same benefits of a specially-designed fixture. Examples include workshop vises, adjustable clamps, and improvised devices such as weights and furniture.

Each component of a fixture is designed for one of two purposes: location or support.

Location

Locating components ensure the geometrical stability of the workpiece. They make sure that the workpiece rests in the correct position and orientation for the operation by addressing and impeding all the degrees of freedom the workpiece possesses.

For locating workpieces, fixtures employ pins (or buttons), clamps, and surfaces. These components ensure that the workpiece is positioned correctly, and remains in the same position throughout the operation. Surfaces provide support for the piece, pins allow for precise location at low surface area expense, and clamps allow for the workpiece to be removed or its position adjusted. Locating pieces tend to be designed and built to very tight specifications.

Support

In designing the locating parts of a fixture, only the direction of forces applied by the operation are considered, and not their magnitude. Locating parts technically support the workpiece, but do not take into account the strength of forces applied by the process and so are usually inadequate to actually secure the workpiece during operation. For this purpose, support components are used.

To secure workpieces and prevent motion during operation, support components primarily use two techniques: positive stops and friction. A positive stop is any immovable component (such as a solid surface or pin) that, by its placement, physically impedes the motion of the workpiece. Support components are more likely to be adjustable than locating components, and normally do not press tightly on the workpiece or provide absolute location.

Support components usually bear the brunt of the forces delivered during the operation. To reduce the chances of failure, support components are usually not also designed as clamps.

Types of Fixtures

Fixtures are usually classified according to the machine for which they were designed. The most common two are milling fixtures and drill fixtures.

Milling Fixtures

Milling operations tend to involve large, straight cuts that produce lots of chips and involve varying force. Locating and supporting areas must usually be large and very sturdy in order to accommodate milling operations; strong clamps are also a requirement. Due to the vibration of the machine, positive stops are preferred over friction for securing the workpiece. For high-volume automated processes, milling fixtures usually involve hydraulic or pneumatic clamps.

Drilling Fixtures

Drilling fixtures cover a wider range of different designs and procedures than milling fixtures. Though workholding for drills is more often provided by jigs, fixtures are also used for drilling operations.

Two common elements of drilling fixtures are the hole and bushing. Holes are often designed into drilling fixtures, to allow space for the drill bit itself to continue through the workpiece without damaging the fixture or drill, or to guide the drill bit to the appropriate point on the workpiece. Bushings are simple bearing sleeves inserted into these holes to protect them and guide the drill bit.

Because drills tend to apply force in only one direction, support components for drilling fixtures may be simpler. If the drill is aligned pointing down, the same support components may compensate for the forces of both the drill and gravity at once. However, though monodirectional, the force applied by drills tends to be concentrated on a very small area. Drilling fixtures must be designed carefully to prevent the workpiece from bending under the force of the drill.

DIE

A die is a specialized tool used in manufacturing industries to cut or shape material mostly using a press. Like molds, dies are generally customized to the item they are used to create. Products made with dies range from simple paper clips to complex pieces used in advanced technology.

Die Forming

Progressive die with scrap strip and stampings.

Forming dies were typically made by tool and die makers and put into production after mounting into a press. The die was a metal block that was used for forming materials like sheet metal and plastic. For the vacuum forming of plastic sheet only a single form was used, typically to form transparent plastic containers (called blister packs) for merchandise. Vacuum forming was considered a simple molding thermoforming process but uses the same principles as die forming. For the forming of sheet metal, such as automobile body parts, two parts may be used: one, called the punch, performed the stretching, bending, and/or blanking operation, while another part that was called the die block securely clamps the workpiece and provided similar stretching, bending, and/or blanking operation. The workpiece may pass through several stages using different tools or operations to obtain the final form. In the case of an automotive component there was usually be a shearing operation after the main forming was done and then additional crimping or rolling operations to ensure that all sharp edges were hidden and to add rigidity to the panel.

Die Components

The main components for die tool sets are:

- Die block: This is the main part that all the other parts are attached to.

- Punch plate: This part holds and supports the different punches in place.

- Blank punch: This part along with the blank die produces the blanked part.

- Pierce punch: This part along with the pierce die removes parts from the blanked finished part.

- Stripper plate: This is used to hold the material down on the blank/pierce die and strip the material off the punches.

- Pilot: This will help to place the sheet accurately for the next stage of operation.

- Guide, back gauge, or finger stop: These parts are all used to make sure that the material being worked on always goes in the same position, within the die, as the last one.

- Setting (stop) block: This part is used to control the depth that the punch goes into the die.

- Blanking dies.

- Pierce die.

- Shank: Used to hold in the presses. It should be aligned and situated at the center of gravity of the plate.

Processes

- Blanking: A blanking die produces a flat piece of material by cutting the desired shape in one operation. The finished part is referred to as a blank. Generally a blanking die may only cut the outside contour of a part, often used for parts with no internal features. Three benefits to die blanking are:

  o Accuracy: A properly sharpened die, with the correct amount of clearance between the punch and die, will produce a part that holds close dimensional tolerances in relationship to the part's edges.

  o Appearance: Since the part is blanked in one operation, the finish edges of the part produces a uniform appearance as opposed to varying degrees of burnishing from multiple operations.

  o Flatness: Due to the even compression of the blanking process, the end result is a flat part that may retain a specific level of flatness for additional manufacturing operations.

- Broaching: The process of removing material through the use of multiple cutting teeth, with each tooth cutting behind the other. A broaching die is often used to remove material from parts that are too thick for shaving.

- Bulging: A bulging die expands the closed end of tube through the use of two types of bulging dies. Similar to the way a chef's hat bulges out at the top from the cylindrical band around the chef's head.

  o Bulging fluid dies: Uses water or oil as a vehicle to expand the part.

  o Bulging rubber dies: Uses a rubber pad or block under pressure to move the wall of a workpiece.

- Coining: Is similar to forming with the main difference being that a coining die may form completely different features on either face of the blank, these features being transferred from the face of the punch or die respectively. The coining die and punch flow the metal by squeezing the blank within a confined area, instead of bending the blank. For example: an Olympic medal that was formed from a coining die may have a flat surface on the back and a raised feature on the front. If the medal was formed (or embossed), the surface on the back would be the reverse image of the front.

- Compound operations: Compound dies perform multiple operations on the part. The compound operation is the act of implementing more than one operation during the press cycle.

- Compound die: A type of die that has the die block (matrix) mounted on a punch plate with perforators in the upper die with the inner punch mounted in the lower die set. An inverted type of blanking die that punches upwards, leaving the part sitting on the lower punch (after being shed from the upper matrix on the press return stroke) instead of blanking the part through. A compound die allows the cutting of internal and external part features on a single press stroke.

- Curling: The curling operation is used to roll the material into a curved shape. A door hinge is an example of a part created by a curling die.

- Cut off: Cut off dies are used to cut off excess material from a finished end of a part or to cut off a predetermined length of material strip for additional operations.

- Drawing: The drawing operation is very similar to the forming operation except that the drawing operation undergoes severe plastic deformation and the material of the part extends around the sides. A metal cup with a detailed feature at the bottom is an example of the difference between formed and drawn. The bottom of the cup was formed while the sides were drawn.

- Extruding: Extruding is the act of severely deforming blanks of metal called slugs into finished parts such as an aluminum I-beam. Extrusion dies use extremely high pressure from the punch to squeeze the metal out into the desired form. The difference between cold forming and extrusion is extruded parts do not take shape of the punch.

- Forming: Forming dies bend the blank along a curved surface. An example of a part that has been formed would be the positive end(+) of a AA battery.

- Cold forming (cold heading): Cold forming is similar to extruding in that it squeezes the blank material but cold forming uses the punch and the die to create the desired form, extruding does not.

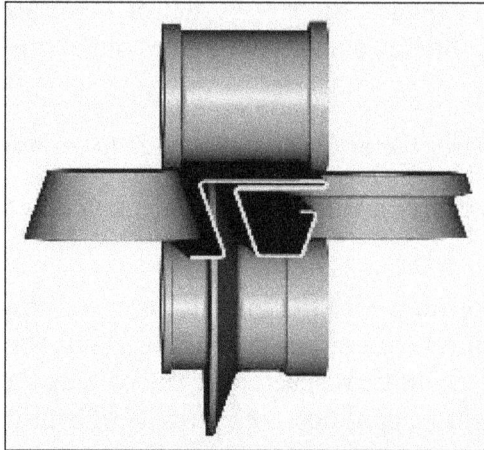

Roll Forming Stand.

- Roll forming: A continuous bending operation in which sheet or strip metal is gradually formed in tandem sets of rollers until the desired cross-sectional configuration is obtained. Roll forming is ideal for producing parts with long lengths or in large quantities.

- Horning: A horning die provides an arbor or horn which the parts are place for secondary operations.

- Hydroforming: Forming of tubular part from simpler tubes with high water pressure.

- Pancake die: A Pancake die is a simple type of manufacturing die that performs blanking and/or piercing. While many dies perform complex procedures simultaneously, a pancake die may only perform one simple procedure with the finished product being removed by hand.

- Piercing: The piercing operation is used to pierce holes in stampings.

- Transfer die: Transfer dies provide different stations for operations to be performed. A common practice is to move the material through the die so it is progressively modified at each station until the final operation ejects a finished part.

- Progressive die: The sheet metal is fed through as a coil strip, and a different operation (such as punching, blanking, and notching) is performed at the same station of the machine with each stroke of a series of punches.

- Shaving: The shaving operation removes a small amount of material from the edges of the part to improve the edges finish or part accuracy. (Compare to Trimming).

- Side cam die: Side cams transform vertical motion from the press ram into horizontal or angular motion.

- Sub press operation: Sub-press dies blank and/or form small watch, clock, and instrument parts.

- Swaging: Swaging (necking) is the process of "necking down" a feature on a part. Swaging is the opposite of bulging as it reduces the size of the part. The end of a shell casing that captures the bullet is an example of swaging.

- Trimming: Trimming dies cut away excess or unwanted irregular features from a part, they are usually the last operation performed.

- Pillar set: Pillar set are used for alignment of dies in press movement.

Steel-rule Die

Steel-rule die, also known as cookie cutter dies, are used for cutting sheet metal and softer materials, such as plastics, wood, cork, felt, fabrics, and paperboard. The cutting surface of the die is the edge of hardened steel strips, known as steel rule. These steel rules are usually located using saw or laser-cut grooves in plywood. The mating die can be a flat piece of hardwood or steel, a male shape that matches the workpiece profile, or it can have a matching groove that allows the rule to nest into. Rubber strips are wedged in with the steel rule to act as the stripper plate; the rubber compresses on the down-stroke and on the up-stroke it pushes the workpiece out of the die. The main advantage of steel-rule dies is the low cost to make them, as compared to solid dies; however, they are not as robust as solid dies, so they are usually only used for short production runs.

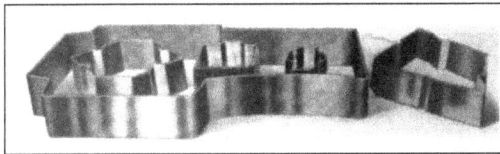

A steel-rule die.   A steel-rule die.

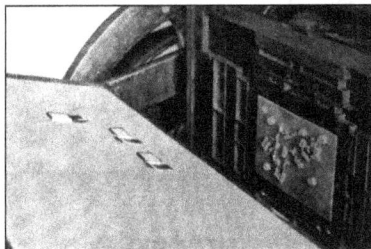

Steel-rule die in a press.

Rotary Die

In the broadest sense, a rotary die is a cylindrical shaped die that may be used in any manufacturing field. However, it most commonly refers to cylindrical shaped dies used to process soft materials, such as paper or cardboard. Two rules are used, cutting and creasing rules. This is for corrugated boards whose thickness is more than 2 mm. Rotary dies are faster than flat dies.

The term also refers to dies used in the roll forming process.

Wire Pulling

Wire-making dies have a hole through the middle of them. A wire or rod of steel, copper, other metals, or alloy enters into one side and is lubricated and reduced in size. The leading tip of the wire is usually pointed in the process. The tip of the wire is then guided into the die and rolled onto a block on the opposite side. The block provides the power to pull the wire through the die.

The die is divided into several different sections. First is an entrance angle that guides the wire into the die. Next is the approach angle, which brings the wire to the nib, which facilitates the reduction. Next is the bearing and the back relief. Lubrication is added at the entrance angle. The lube can be in powdered soap form. If the lubricant is soap, the friction of the drawing of wire heats the soap to liquid form and coats the wire. The wire should never actually come in contact with the die. A thin coat of lubricant should prevent the metal to metal contact.

For pulling a substantial rod down to a fine wire a series of several dies is used to obtain progressive reduction of diameter in stages.

Standard wire gauges used to refer to the number of dies through which the wire had been pulled. Thus, a higher-numbered wire gauge meant a thinner wire. Typical telephone wires were 22-gauge, while main power cables might be 3- or 4-gauge.

MATERIAL-HANDLING EQUIPMENT

Material handling equipment encompasses a diverse range of tools, vehicles, storage units, appliances and accessories involved in transporting, storing, controlling, enumerating and protecting products at any stage of manufacturing, distribution consumption or disposal.

Types of Material Handling Equipment

The four main categories of material handling equipment include storage, engineered systems, industrial trucks, and bulk material handling.

Storage and Handling Equipment

Storage equipment is usually limited to non-automated examples, which are grouped in with engineered systems. Storage equipment is used to hold or buffer materials during "downtimes," or times when they are not being transported. These periods could refer to temporary pauses during long-term transportation or long-term storage designed to allow the buildup of stock. The majority of storage equipment refers to pallets, shelves or racks onto which materials may be stacked in an orderly manner to await transportation or consumption. Many companies have investigated increased efficiency possibilities in storage equipment by designing proprietary packaging that allows materials or products of a certain type to conserve space while in inventory.

Examples of storage and handling equipment include:

- Racks, such as pallet racks, drive-through or drive-in racks, push-back racks, and sliding racks, are a basic but important method of storage, saving floor space while keeping their contents accessible.

- Stacking frames are stackable like blocks, as their name implies. They allow crushable pallets of inventory, such as containers of liquid, to be stacked to save space without damage.

- Shelves, bins, and drawers. Shelves, another basic storage method, are less open than racks. Used with bins and drawers, they're more able to keep smaller and more difficult to manage materials and products stored and organized. Shelving types can include boltless, cantilever, revolving, and tie-down.

- Mezzanines, a type of indoor platform, help to create more floor space in a warehouse or other storage building for offices or more storage. Typical types include modular, movable, rack supported, building supported, and free-standing versions.

Engineered Systems

Engineered systems cover a variety of units that work cohesively to enable storage and transportation. They are often automated. A good example of an engineered system is an Automated Storage and Retrieval System, often abbreviated AS/RS, which is a large automated organizational structure involving racks, aisles and shelves accessible by a "shuttle" system of retrieval. The shuttle system is a mechanized cherry picker that can be used by a worker or can perform fully automated functions to quickly locate a storage item's location and quickly retrieve it for other uses.

Other types of engineered systems include:

- Conveyor systems come in a variety of types, depending on what they are meant to transport, including vibrating, overhead, chain, vertical, and apron conveyors.

- Automatic Guided Vehicles (AGV) are independent computer-operated trucks that transport loads along a predetermined path, with sensors and detectors to avoid bumping into anything.

Industrial Material Handling Trucks

Industrial trucks (material handling trucks) refer to the different kinds of transportation items and vehicles used to move materials and products in materials handling. These transportation devices can include small hand-operated trucks, pallet jacks, and various kinds of forklifts. These trucks have a variety of characteristics to make them suitable for different operations. Some trucks have forks, as in a forklift, or a flat surface with which to lift items, while some trucks require a separate piece of equipment for loading. Trucks can also be manual or powered lift and operation can be walk or ride, requiring a user to manually push them or to ride along on the truck. A stack truck can be used to stack items, while a non-stack truck is typically used for transportation and not for loading.

There are many types of industrial trucks:

- Hand trucks, one of the most basic pieces of material handling equipment, feature a small platform to set the edge of a heavy object on, and a long handle to use for leverage. Whatever is being moved must be tipped so that it rests on the handle, and is carried at a tilt to its destination.

- Pallet trucks, also known as pallet jacks, are a type of truck specifically for pallets. They slide into a pallet and lift it up to move it. Pallet trucks come in both manual and electrical types.

- Walkie Stackers transport and lift pallets like a forklift, though they don't include a place for the operator to ride in. They come in both powered or manual versions.

- Platform trucks are hand trucks low to the ground, with a wide platform for transporting goods.

- Order pickers lift the operator several feet above the ground on a platform so they can retrieve or store goods on high shelves.

- Sideloaders, also known as VNA (Very Narrow Aisle) trucks, are meant to fit in narrow warehouse aisles, as they can load objects from different directions. They're also good for long, awkward products that need moving.

- Many types of AGV, or automatic guided vehicles, shuttle products along a route automatically, without human guidance.

Bulk Material Handling Equipment

Bulk material handling refers to the storing, transportation and control of materials in loose bulk form. These materials can include food, liquid, or minerals, among others. Generally, these pieces of equipment deal with the items in loose form, such as conveyor belts or elevators designed to move large quantities of material, or in packaged form, through the use of drums and hoppers.

- Conveyors, as mentioned above, come in a wide variety of types for different types of bulk material.

- Stackers, which are usually automated, pile bulk material onto stockpiles, moving between two points along rails in a yard.

- Reclaimers are the opposite of stackers, retrieving materials from stockpiles, some using bucket wheels to carry the material while others are scraper or portal style.

- Bucket elevators, also known as grain legs, use buckets attached to a rotating chain or belt to carry material vertically.

- Grain elevators are tall buildings specifically for storing grain. They include equipment to convey the grain to the top of the elevator, where it is sent out for processing.

- Hoppers are funnel-shaped containers that allow material to be poured or dumped from one container to another. Unlike a funnel, though, hoppers can hold material until it's needed, then release it.

- Silos are generally large storage structures for bulk materials, though they don't necessarily include equipment to convey the material to the top of the structure like grain elevators. Different varieties include tower, bunker, and bag silos.

Lean Manufacturing Tools

Lean manufacturing tools are methods to help create a Lean environment and achieve goals around reduced waste, improved efficiency and increased customer value. The following list covers our top ten (of many) Lean manufacturing tools.

PDCA Problem Solving Cycle

PDCA stands for Plan, Do, Check, Act and offers a visual way to represent a typical problem solving cycle. It covers planning for a specific goal, doing the work required by that plan, checking the results of the work and acting to fix any unsatisfactory results. Like many Lean manufacturing tools, PDCA focuses on identifying and solving problems quickly. It also helps everyone involved see the impact of their role on the end product delivered to customers.

The Five Whys

The Five Whys is another one of several Lean manufacturing tools used to identify the root cause of a problem. Quite simply, it requires participants to continually ask "why?" questions (typically five or fewer times) to peel back the layers. This questioning allows teams to diagnose problems without any statistical analysis and often identifies multiple root causes and the relationships between them.

Continuous Flow (aka One Piece Flow)

Continuous Flow is a Lean manufacturing tool that calls on teams to manufacture smaller batches. That's because smaller batches can go through the production system faster and because having smaller batches allows teams to regularly examine the output to make any necessary improvements for the next batch, thereby eliminating waste.

Cellular Manufacturing

Cellular manufacturing supports continuous flow by calling on teams to arrange workstations based on the parts they produce in order to minimize travel time for those parts and allow for rapid feedback across stations about any issues. Together, these two Lean manufacturing tools also enable teams to produce smaller, more efficient batches. Organizations typically achieve cellular manufacturing by arranging workstations in a "U" formation.

Five S

The Five S is another of the Lean manufacturing tools focused on workstations. Specifically, it dictates how teams should organize materials and keep workstations cLean to maximize efficiency. The five S's are: Sort (remove any unnecessary materials), set in order (arrange materials so they are easy to find and access), shine (cLean the workspace regularly), standardize (make the previous three S's a standard routine) and sustain (institute regular audits).

Total Productive Maintenance (TPM)

Total productive maintenance is a Lean manufacturing tool that emphasizes operational efficiency for equipment and safety for workers. Building on the Five S approach, TPM asks workers to help maintain their equipment to avoid accidents, breakdowns, defects and delays. Part of TPM is the Overall Equipment Effectiveness (OEE) metric, which measures the percentage of working time that is actually productive. OEE is based on scores for availability, performance speed and output quality. Multiplying these three numbers gives the OEE metric, with 85% best-in-class for organizations using Lean manufacturing tools, 60% average for organizations using Lean manufacturing tools and 40% average for organizations not embracing any Lean manufacturing tools.

Takt Time

Takt time is one of the Lean manufacturing tools focused on customer value. It measures the average rate at which teams must manufacture products to meet demand. To calculate takt time, divide the working time available for production (in hours, days, weeks) by the units required to meet customer demand. Using this Lean manufacturing tool, if a team works 40 hours a week and the company expects customers to buy 80 units a week, the takt time is 0.5 (40 / 80). This means the team must produce one unit in half an hour to meet customer demand.

Standardized Work

Standardized work and takt time are two Lean manufacturing tools that work together. Specifically, standardized work helps teams achieve their takt time by creating and documenting a set, repeatable process for how teams should function. For example, it documents the steps teams should take, the materials they need and the time required for each step.

Mistake Proofing

Mistake proofing focuses on detecting mistakes as they occur (either automatically through technology or manually through inspection) and notifying workers accordingly. Like all Lean manufacturing tools, it helps eliminate waste and increase efficiency, this time by surfacing errors as they occur to prevent defective parts from moving further down the process.

Leveling the Workload

Leveling the workload calls for teams to manufacture products consistently despite inconsistencies in customer demand. That means even though customers might buy 100 units one week and 60 units the next week, teams should maintain a consistent output over a set period of time (e.g. a week or a month). This consistency helps teams remain efficient since they do not need to switch setups to manufacture different products in unpredictable patterns.

References

- Vincent Wang, Xi; Xu, Xun W. (2013-08-01). "An interoperable solution for Cloud manufacturing". Robotics and Computer-Integrated Manufacturing. 29 (4): 232–247. Doi:10.1016/j.rcim.2013.01.005

- What-is-digital-manufacturing: ibaset.com, Retrieved 20 January, 2019

- Hermann Kühnle (2010). Distributed Manufacturing: Paradigm, Concepts, Solutions and Examples. Springer. ISBN 978-1-84882-707-3. Retrieved 7 May 2013

- Laser-additive-manufacturing, engineering, topics: sciencedirect.com, Retrieved 21 February, 2019

- Y., Lu; X. Xu; J. Xu (2014). "Development of a Hybrid Manufacturing Cloud". Journal of Manufacturing Systems. 33 (4): 551–566. Doi:10.1016/j.jmsy.2014.05.003

- Introduction-to-Cyber-Manufacturing: researchgate.net, Retrieved 22 March, 2019

- Wu, D.; Thames, J.L.; Rosen, D.W.; Schaefer, D. (2013). "Enhancing the Product Realization Process with Cloud-Based Design and Manufacturing Systems". Journal of Computing and Information Science in Engineering. 13 (4): 041004. Doi:10.1115/1.4025257

- Digital-Factory-Theory-and-Practice: researchgate.net, Retrieved 23 April, 2019

- Felix Bopp (2010). Future Business Models by Additive Manufacturing. Verlag. ISBN 978-3836685085. Retrieved 4 July 2014

Digital Manufacturing

A consolidated approach to manufacturing which primarily makes use of computing systems is defined as digital manufacturing. Computer-integrated manufacturing, digital factory, cloud manufacturing, cyber manufacturing, digital materialization, distributed manufacturing, cloud-based designing, laser rapid manufacturing, etc. are some of its applications. All these diverse principles and applications of digital manufacturing have been carefully analyzed in this chapter.

Digital manufacturing is the application of digital technologies to manufacturing. It is all about having the right information, at the right place, at the right time. The goal is to link disparate systems and span processes across all departments and functions within the value chain. By doing so, the entire product lifecycle is impacted – from design to production to servicing of the final products. With digital manufacturing systems, each stakeholder gains quicker access to more accurate data. This improves process efficiency and heightens the quality of organizational decision-making.

Digital Manufacturing Connects Processes

Digital manufacturing requires integration between PLM, ERP, shop floor applications and equipment to enable the exchange of product-related information between digital design and physical manufacturing execution. Manufacturers can achieve time-to-market and volume goals, as well as realize cost savings by establishing a digital thread that provides a formal framework for digital exchange that can analyze data throughout the product lifecycle, transforming it into actionable information. The digital thread integrates as-designed requirements, validation and inspection records, as-built data, as-flown data, and as-maintained data.

Smart, connected products can send customer data to product managers to help anticipate demand and meet ongoing maintenance needs. The result is better designed products maintained in detail throughout the product life cycle. By involving customers during the entire product lifecycle, requirements can be fulfilled faster and with far fewer iterations.

Three Dimensions of Digital Manufacturing

Digital manufacturing can be broken down into three dimensions: (a) Product Life Cycle, (b) Smart Factory, and (c) Value Chain Management.

The Product Life Cycle starts with an engineering design definition and follows through sourcing, production and service life. Digital data for each step includes every incorporated revision, any approved deviations from design specifications and how these are executed across the lifecycle.

The Smart Factory is all about automation. It encompasses smart machines, sensors and tooling to provide workers with real-time data about the processes they are executing. It forms the bridge between Operations Technology (OT) that exchanges data directly with machines and tooling, and Information Technology (IT) systems and apps. Both are enhanced by business intelligence systems that perform in-depth analysis. This leads to real-time visibility of factory processes, process control optimization, and insight into potential areas of performance or process improvement.

Value Chain Management focuses on minimizing resources and accessing value at each stakeholder function along the chain. It results in optimal process integration, decreased inventories, better products, and enhanced customer satisfaction.

Benefits of Digital Manufacturing

Digital manufacturing brings together complex manufacturing processes across departments, and eliminates paper processes that can be fraught with errors and repeated information. Benefits include:

- Increased efficiency through automated exchange of data.
- Avoidance of costly errors due to missed or misinterpreted data.
- Quicker turnaround at all levels of the value chain.
- Greater insight at critical decision points.
- Real-time visibility into the effects of changes to processes, equipment, systems or components.
- Faster pace of innovation.
- Lowered cost of production and maintenance.

Digital Manufacturing in Aerospace and Defense

Some areas of the aerospace-and-defense (A&D) industry are deploying digital tools to integrate their enormously complex supply networks. A modern jet turbine engine has hundreds of individual parts, for example, some made in-house and others sourced from dozens of vendors. Through digital manufacturing, cloud computing-based tools allow suppliers to collaborate efficiently. This greatly reduces the labor required to manage design changes and minimizes risk across the supply network. Boeing is a good example of an organization realizing the benefits of digital manufacturing. The A&D giant developed its 777 and 787 airframes using all-virtual design. This cut time to market by more than 50 percent.

The way people and organizations use information is shifting dramatically. As the intensity of global competition raises the stakes, the pressure is on manufacturers to think differently about business models while still cultivating additional revenue streams and finding ways to outflank the competition.

COMPUTER-INTEGRATED MANUFACTURING

Computer-integrated manufacturing (CIM) is the manufacturing approach of using computers to control entire production process. This integration allows individual processes to exchange information with each other and initiate actions. Although manufacturing can be faster and less error-prone by the integration of computers, the main advantage is the ability to create automated manufacturing processes. Typically CIM relies of closed-loop control processes, based on real-time input from sensors. It is also known as flexible design and manufacturing.

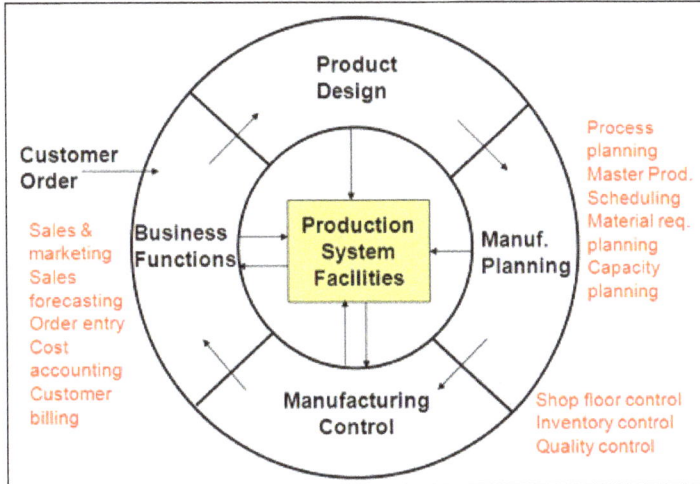

Manufacturing Systems Integration Program.

Computer-integrated manufacturing is used in automotive, aviation, space, and ship building industries. The term "computer-integrated manufacturing" is both a method of manufacturing and the name of a computer-automated system in which individual engineering, production, marketing, and support functions of a manufacturing enterprise are organized. In a CIM system functional areas such as design, analysis, planning, purchasing, cost accounting, inventory control, and distribution are linked through the computer with factory floor functions such as materials handling and management, providing direct control and monitoring of all the operations.

As a method of manufacturing, three components distinguish CIM from other manufacturing methodologies:

- Means for data storage, retrieval, manipulation and presentation;

- Mechanisms for sensing state and modifying processes;

- Algorithms for uniting the data processing component with the sensor/modification component.

CIM is an example of the implementation of information and communication technologies (ICTs) in manufacturing.

CIM implies that there are at least two computers exchanging information, e.g. the controller of an arm robot and a micro-controller of a [[]].

Some factors involved when considering a CIM implementation are the production volume, the experience of the company or personnel to make the integration, the level of the integration into the product itself and the integration of the production processes. CIM is most useful where a high level of ICT is used in the company or facility, such as CAD/CAM systems, the availability of process planning and its data.

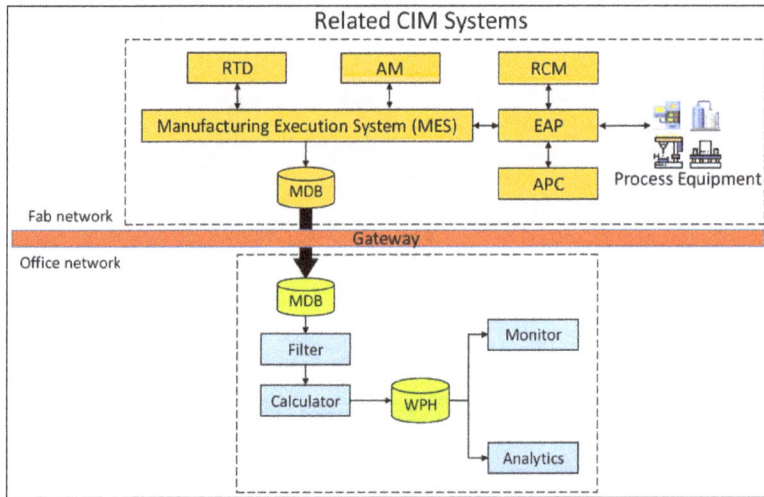

CIM & production control system: Computer Integrated Manufacturing is used to describe
the complete automation of a manufacturing plant, with all processes running under computer control
and digital information tying them together.

Key Challenges

There are three major challenges to development of a smoothly operating computer-integrated manufacturing system:

- Integration of components from different suppliers: When different machines, such as CNC, conveyors and robots, are using different communications protocols (In the case of AGVs, even differing lengths of time for charging the batteries) may cause problems.

- Data integrity: The higher the degree of automation, the more critical is the integrity of the data used to control the machines. While the CIM system saves on labor of operating the machines, it requires extra human labor in ensuring that there are proper safeguards for the data signals that are used to control the machines.

- Process control: Computers may be used to assist the human operators of the manufacturing facility, but there must always be a competent engineer on hand to handle circumstances which could not be foreseen by the designers of the control software.

Subsystems

A computer-integrated manufacturing system is not the same as a "lights-out factory", which would run completely independent of human intervention, although it is a big step in that direction. Part of the system involves flexible manufacturing, where the factory can be quickly modified to produce different products, or where the volume of products can be changed quickly with the aid of computers. Some or all of the following subsystems may be found in a CIM operation:

Computer-aided techniques:

- CAD (computer-aided design).

- CAE (computer-aided engineering).

- CAM (computer-aided manufacturing).

- CAPP (computer-aided process planning).

- CAQ (computer-aided quality assurance).

- PPC (production planning and control).

- ERP (enterprise resource planning).

- A business system integrated by a common database.

Devices and equipment required:

- CNC, Computer numerical controlled machine tools.

- DNC, Direct numerical control machine tools.

- PLCs, Programmable logic controllers.

- Robotics.

- Computers.

- Software.

- Controllers.

- Networks.

- Interfacing.

- Monitoring equipment.

Technologies:

- FMS, (flexible manufacturing system).

- ASRS, automated storage and retrieval system.

- AGV, automated guided vehicle.

- Robotics.

- Automated conveyance systems.

Others:

- Lean manufacturing.

CIMOSA

CIMOSA (Computer Integrated Manufacturing Open System Architecture), is a 1990s European proposal for an open systems architecture for CIM developed by the AMICE Consortium as a series of ESPRIT projects. The goal of CIMOSA was "to help companies to manage change and integrate their facilities and operations to face world wide competition. It provides a consistent architectural framework for both enterprise modeling and enterprise integration as required in CIM environments".

CIMOSA provides a solution for business integration with four types of products:

- The CIMOSA Enterprise Modeling Framework, which provides a reference architecture for enterprise architecture.

- CIMOSA IIS, a standard for physical and application integration.

- CIMOSA Systems Life Cycle, is a life cycle model for CIM development and deployment.

- Inputs to standardization, basics for international standard development.

CIMOSA according to Vernadat (1996), coined the term business process and introduced the process-based approach for integrated enterprise modeling based on a cross-boundaries approach, which opposed to traditional function or activity-based approaches. With CIMOSA also the concept of an "Open System Architecture" (OSA) for CIM was introduced, which was designed to be vendor-independent, and constructed with standardised CIM modules. Here to the OSA is "described in terms of their function, information, resource, and organizational aspects. This should be designed with structured engineering methods and made operational in a modular and evolutionary architecture for operational use".

Application

There are multiple areas of usage:

- In Industrial and Production engineering.

- In mechanical engineering.

- In electronic design automation (printed circuit board (PCB) and integrated circuit design data for manufacturing).

DIGITAL FACTORY

The digital economy represents the pervasive use of IT (hardware, software, applications and telecommunications) in all aspects of the economy, including internal operations of organizations (business, government and non-profit); transactions between organizations; and transactions between individuals, acting both as consumers and citizens, and organizations.

Future production environment will require holonic approach of integration of information technologies (IT) into production area. On one side it is represented through IT applications in a real machinery and equipment. On the other side IT has to be employed in the design of production systems as well.

IT has been the key factor responsible for reversing the 20-year productivity slowdown from the mid-1970s to the mid-1990s and in driving today's robust productivity growth.

Any year automotive exhibitions show new models, merry-go-round of innovations turns faster and faster. Original Equipment Manufacturers (OEMs) introduce any 2 to 3 months new models which very often requires changes of production processes. Extremely short innovation cycles and products customisation significantly change all industries. The innovation is successful only in case, that it is quickly launched on market. The collaboration of single partners is very important not only in product development but in production planning and control too. There exists a lot of chaos and supplementary costs by the launching of new products. These supplementary costs often reach millions of Euros.

Digital Factory is the phenomenon having its background in computer aided and computer integrated technologies and advanced virtual reality technologies. This phenomenon became very important mainly at the beginning of 21.Century.

The continually changing conditions of global markets combined with customer behaviour changes brought strong requirements for current producers e.g.: short time to markets, high quality, low cost, short production throughput times, etc. All these changes demand quick response with high orders fulfilment security.

Current markets and customer's environment require from designers of manufacturing systems the utilisation of new advanced approaches and tools in the design of future manufacturing systems. Digital Factory seems to be one of the most appropriate approaches to fulfil this task offering all required functions e.g.: central database, digital models, integrated data management, modelling and simulation functions, visualisation through virtual reality etc. Virtual reality became a common tool used in Digital Factory environment. Digital Factory creates the environment for digital innovation of any part of production systems, e.g. products, processes and resources.

Virtual Reality is a computer technology, which supported by hardware and software, enables to create virtual models of objective reality and use them for the generation of perceive feelings of people.

Virtual Reality technologies are possible to use for design of 3D spatial models as well as for 3D modelling and examination of properties of real objects (Medvecký, Š. , 2007). On the other side Virtual Reality enables to create „real" spatial environment, in which the man can conduct required activities. The possibilities for development of Virtual Reality technologies are tremendous and they still grow.

Digital Factory entitles the virtual environment for the lifecycle design of manufacturing processes and manufacturing systems using simulation and virtual reality technologies to optimize performance, productivity, timing, costs and ergonomics. There exists a comprehensive source of materials defining and describing Digital Factory.

Digital Factory environment uses 3D digital models (DMUs) and associated information for visualizing, modelling, and simulating production processes and production systems with the target of effective and productive real production within resource constraints. It enables to design, analyse and predict the future behaviour of designed production systems supported by computer simulation. Computer simulation plays very important role in the study of behaviour of real and artificial systems, almost in any scientific area.

Digital Factory (DF) represents a virtual picture of a real production. It is the environment integrated by computer and information technologies, in which the reality is replaced by virtual computer models. Such virtual solutions enable to verify all conflict situations before real implementation of factories and to design optimised solutions.

Product Lifecycle Management (PLM) is a business strategy supporting companies in product data sharing, and leveraging of corporate knowledge for the development of products for their lifecycle. PLM enables to operate and manage the entire network of all players (enterprise, suppliers, customers) as a single entity. Nowadays, the need for "Scientific Management" as proposed by Taylor has been extended to smart Product Life Cycle Management.

Different types of software are linked in PLM solutions, which control different parts of the manufacturing cycle. Computer Aided Design (CAD) systems define what will be produced. Computer Aided Engineering (CAE) defines production processes and systems required for product´s production. Computer Aided Manufacturing (CAM) and Manufacturing Process Management (MPM) define how it is to be built. ERP answers when and where it is built. Manufacturing Execution System (MES) provides shop floor control and simultaneously manufacturing feedback. The storing of information digitally aids communication, but also removes human error from the design and manufacture process. Computer Integrated Manufacturing (CIM) and Product Data Management (PDM) were recently replaced by term Digital Manufacturing (DM) which currently is conceptually very close to Digital Factory.

Digital Factory represents integration chain between CAD systems and ERP solutions, as it is shown in figure.

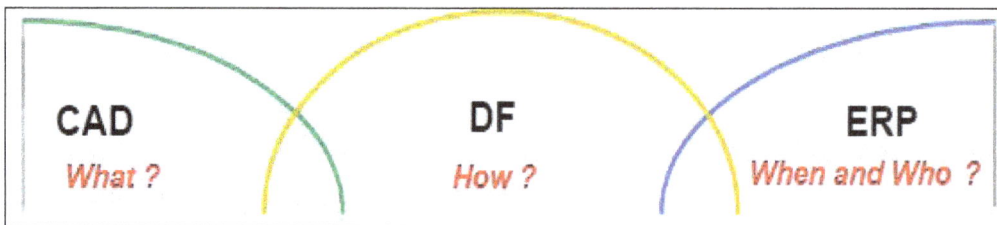

Integration of Information Systems in Production.

One of very important properties of Digital Factory is the vision to realize process planning and product development with parallel utilisation of common data. It is very important to gain all required data only one time and then to manage them with the uniform data control, so that all software systems will be able to utilize it. The integration is one of the main conditions for the implementation of Digital Factory.

Digital Factory integrates three main elements:

- Digital product, with its static and dynamic properties,
- Digital production planning,
- Digital production, with the possibility of utilisation of planning data for enterprise processes effectiveness growth.

Key Enterprise Processes

Digital Factory concept prioritizes the six most significant areas, according to their influence on production process flow. Any area covers the set of tools which all together integrate the whole production process, from product design to its production:

- Product design systems (including modelling and simulation).

- Process planning systems (process and production plans, assembly plans, welding plans, tools, jigs, work standardization, value analysis, cost analysis, etc.).

- Production process detail and validation systems (NC production process simulation, assembly, inspection, maintenance, production operations etc.). The utilisation of process plans, graphs and special BOMs which offer clear view about relationships between processes and resources already in conceptual design phase.

- Production engineering systems (complex production scenarios, layout, industrial engineering, time analysis, ergonomics analysis, design and analysis of production and assembly systems, loading of machines, determination and optimisation of workers loading, etc.).

- Production planning and control systems (ERP planning systems, scheduling, pull control, levelled production, mixed production, etc.).

- Automation and process control systems (automatic generation of control programmes for control and monitoring of automated production systems, PLC, industrial robots, etc.).

Digital Factory Implementation Methodology

Rough procedure of Digital Factory implementation is as follows:

- Definition of total standards and production principles for entire planning operations, creation of primitives and customer databases.

- First data collection and organisation with the utilisation of data management system. All responsible persons have direct access to the data, its addition, inspection and changes.

- In this phase, Digital Factory system improves co-ordination and synchronisation of individual processes throughout their "networking" supported by workflow management system.

In the fourth phase, Digital Factory system takes automatically some routine and checking activities, which are very time consuming in common systems. Implemented system insures high quality of all outputs.

Digital Factory Application Area

Digital Factory is appropriate mainly as a support for the batch manufacturing of highly sophisticated products, their planning, simulation and optimisation. Its main current application area is automotive industry, mechanical engineering industry, aerospace and ship building industry as well as electronics and consumer goods industries. 3D digital model of products (DMU) creates currently basic object for the work in digital manufacturing environment. There exists possibility to optimise products, processes and production systems even during the development phase with the utilisation of 3D visualisation and modelling techniques. Such solution brings time to market reduction and significant cost reduction.

Cutting process simulation enables generation of real NC-part programmes for chosen production machines. The visualisation of machines and operational handling processes enables to reveal shortages in design of machines and to realise changes, remove errors and increase their effectiveness before real utilisation.

The system for the design of shop floor 3D layouts and generation of 3D models of production halls is missing in current Digital Factory solutions. It is possible to create the 3D model of production hall directly in CAD systems. Such solution is advantageous by new layouts or by new production systems designs. But, production halls do exists, in majority of real cases. By such conditions, it is often more effective to create 3D model of production hall with the utilisation of Reverse Engineering technologies and 3D scanners.

The material flow simulation enables to optimise the movement of material, to reduce inventories and to support value added activities in internal logistics chain.

The subsystems for effective ergonomics analysis utilise international standards as NIOSH, RULA, etc., which enable right planning and verification of man-machine interactions on the single workplaces.

The highest level of analysis represents computer simulation of production and robotics systems which enables optimisation of material, information, value and financial flows in the factory.

Simulation and Emulation Production Environment

ZIMS – Innovative Environment.

As a part of Digital Factory concept, a comprehensive holistic solution for the design and testing of new production systems is currently being developed, known as ZIMS - Zilina Intelligent Manufacturing System. ZIMS represents a mixed virtual and real environment which enables to design and develop advanced intelligent solutions for industry. Figure shows picture of current ZIMS state. It integrates many enterprise areas like: product design, Rapid Prototyping, product properties simulation and testing, new technology design, layout design and optimization, material flow optimization, handling, production control, ergonomics, simulation, etc. The computer simulation plays a central role in ZIMS concept, beginning from product design up to manufacturing processes simulation. Such simulation environment is known as Enterprise Simulation Management.

Chosen Examples from Research and Industry

The following part presents the chosen results of the research studies conducted in co- operation with the industrial OEM producers. As the results of research projects showed the Digital Factory solutions can save billions of Euros in industry.

The creative thinking methods became very popular in the design of new products. The methods of creative problems solution based on contradictions, like "general methodologies": TRIZ, WOIS, CREAX, DIVA, etc., are very often used by the design of new products.

The following example shows the application of Rapid Prototyping and Vacuum Casting technologies to the development and production of gear boxes.

The technologies of Rapid Prototyping are appropriate especially in case the company has to offer a quick response to the customer requirements on innovation. The automotive industry requires fast response to customer requirements.

The DMU model (DMU) of given gearbox was created by Reverse Engineering technologies (Minolta 3D Laser Scanner VI 900). Figure shows the real VW Gearbox and its scanned clouds of point's model.

Real VW Gearbox & Its Scan.

The 3D model of real gearbox was generated in Catia and FDM (Fused Deposition Modelling) prototype was produced through Rapid Prototyping.

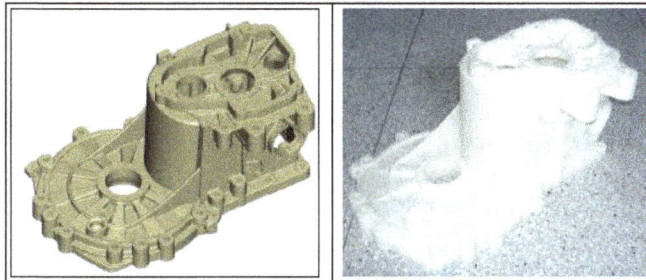

3D Model in Catia and FDM Prototype.

Laser Scanning for Quick Digitization of Real Objects

Mainly classical approaches are being used for digitisation and geometric analyses of the existing production systems. Information about the real state of the production system is, in case of complex production systems, obtained using the measuring tape, or laser measurers. Using such approach makes digitization of the whole enterprise extremely time-demanding and expensive. It is also a potential source of waste, inaccuracies and errors. It is much faster, much more effective and qualitatively better to create the 3D models of the existing production systems using the 3D laser scanners. These make possible to transform the existing, real 3D word, into its exact 3D digital copy which correctly reproduces the exact geometry of the recorded space and can simply be used for any computer analyses, in a matter of a few moments.

Thus obtained 3D digital model (so-called master model) can be used in all designer professions; it can be used by analysts as well as by the factory's management. Using the internet it is possible to share such model from anywhere worldwide. Its accessibility makes it easier to eliminate errors. Designers from all over the world can simultaneously work on new projects without any need to travel on to the spot and manually do all the measurements required before they start to design.

Extensive research is currently underway, all over the world, in the sphere of utilising the digital methods for digitization, modelling, analysing, simulation, recording and presenting of real objects.

The sphere of creating, modelling and storing 3D digitalized virtual models of real objects is one of the most significant spheres, which are able to radically influence the effectiveness of producers. Research and development in this high-tech sphere is technically and financially demanding. The most significant automotive and electronics companies are well aware of the permanent need to innovate their products, which is why they release a new model every 2-3 months. Innovation can only be successful if it is swiftly put on the market. To fulfil the requirement to shorten the whole production cycle of a product from its design to delivering it to the customer keeping the costs as low as possible is the most important prerequisite of success of every enterprise. The launch of a new product is always connected with the initial chaos, which increases the realisation costs behindhand.

The system for the creation of 3D production layouts and the generation of DMUs of production halls or FMUs is what Digital Factory solutions miss today. It is principally possible to design the DMU of production halls and production layouts using the direct CAD system approach. Such solution is convenient when designing new production systems. However, the more frequent case is that the production halls do already exist. That is the reason why it is often more effective to create production hall DMU using the Reverse Engineering technologies (e.g. 3D laser scanning.).

Reverse Engineering is the step needed to take to be able to achieve high efficiency and accuracy of digitization, not only considering the existing equipment, but also when the production layout themselves come into question. It opens up new opportunities to realize virtual designing. Creation of 3D-DMU of large objects using the 3D scanning is, at the moment, the joining link between virtual reality and real virtuality.

The 3D laser scanning (3D-LS) is one of the Reverse Engineering technologies which are usually used for the digitization of real objects. The digitization through 3D laser scanning represents one of the most productive and effective ways of how to get the high quality 3D digital models of current production systems. Those technologies enable simultaneously the transformation of current real worlds into their 3D clouds of points copies which represents their spatial geometry. Such 3D models are very useful for the analysis of current production systems. The 3D-LS technologies became a part of the Digital Factory technologies. Their advantage lays in the simplicity and a huge potential of cost and time savings by the development of 3D models of current objects.

Based on a research by 3D laser scanners users were achieved following costs savings.

- According to the research by customer's consistent application of 3D factory it can save 30-40 % additional costs and time in projects.

- Complex 3D data are basis for detection and elimination of clash causes. It can be saved up to 2 % of investment costs by investment in to factories by using detection and elimination clash causes.

- Created and complex 3D DMU allows accurate, quick, easy and effective change management. Time in this case is featured, so these planning and management systems are also marked as 3D-CAD-Planning tools (also marked 4D). The utilization of automatic scanning based on ahead set plan allows quicker to obtain a new and real 3D DMU. The planning system on the other side allows with one click to realize changes in integrated form, which were in the past solved by groups of specialists in months.

The chosen Reverse Engineering technologies were tested in the framework of co-operation of the University of Žilina and Thyssen Krupp – PSL and Whirlpool companies. Figure shows the DMU model of handling device developed using 3D laser scanning and modelling in Catia

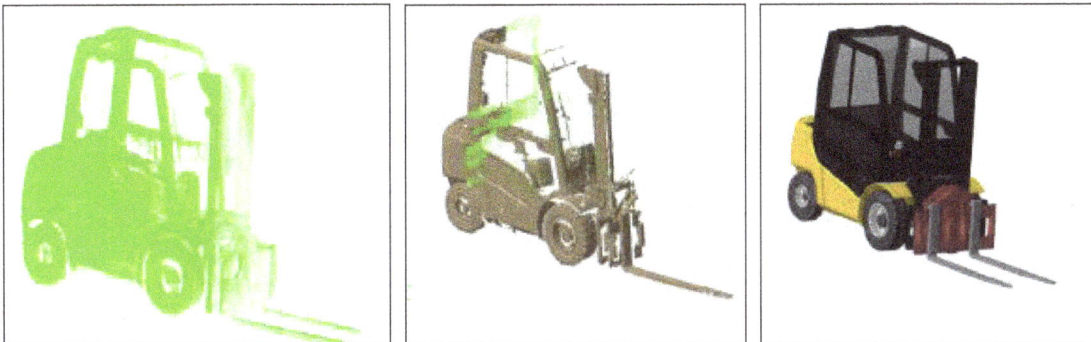

Clouds of Point Model.   Transfer in CAD.   3D Digital Model.
DMU Model of Handling Device.

The 3D virtual models of production halls were developed using 3D laser scanning (Faro LS 880 HE) and modelling in Bentley HLS. The following example shows the result of production halls DMUs development through 3D laser scanning and modelling, including detailed energy networks and transportation systems.

DMUs of Production Hall – VW Slovakia.

The designer of manufacturing systems can use the existing 2D production hall model and to develop new 3D digital model in CAD system. It is often advantageous to generate the 3D model on the basis of existing 2D model platform. A comprehensive research was done, in co-operation with Thyssen Krupp – PSL, in which the complex digital cycle was conducted. After 3D model of production hall was Developed in Bentley HLS system, based on 3D laser scans, all objects (e.g. production hall, machine tools, handling devices, etc.) were integrated into 3D model and through this approach the comprehensive DMU was developed. Then the new assembly concept with optimized material flow was developed. Consequently 3D simulation model of complex production system was developed and used by optimization of control of the designed production system. The final solution represents a comprehensive model and the research on this area continues. Figure shows the part of FMU of Thyssen Krupp – PSL factory.

DMU Model of Manufacturing and Assembly – Thyssen Krupp-PSL.

The Development of an Assembly Line Digital Model

The DMU model of a real gearbox was developed using Reverse Engineering technology (3D laser scanning), in the framework of co-operation in research with VW Slovakia. The developed gearbox DMU was used for the development of the digital model of the entire gearboxes assembly line. Figure shows the real VW gearbox and its digital model (DMU).

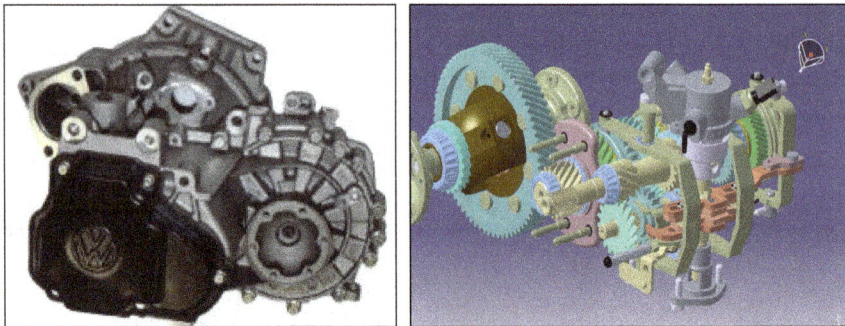

Real Versus Virtual VW Gearbox.

Based on the gearbox DMU and a real assembly system a set of DMUs of VW production workplaces and transportation equipments was developed.

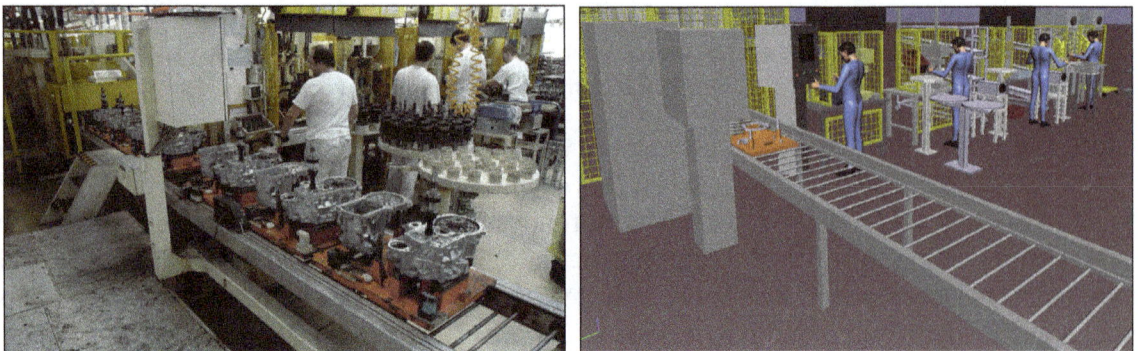

DMUs of Assembly Workplace.

The design of workplaces was especially checked by ergonomics analysis whereas manikin concept of Delmia V5 Human was used. The final solution of designed workplaces was tested through the static simulation /animation.

Static Digital Model of Assembly Line.

The static virtual model of a given gearbox assembly line was developed through integration of individual DMUs into manufacturing system scene. The dynamics of the real assembly system was checked in the 3D simulation environment Quest. The set of simulation experiments was conducted with the developed simulation model which showed bottlenecks stations and the possibilities for performance improvement of gearbox assembly line. Afterwards an FMU of the whole assembly line for gearboxes assembly in VW Slovakia was developed. This FMU represents the complex digital model of the entire VW assembly line.

Simulation and Emulation System

Central European Institute of Technology (CEIT) has long time been developing the simulation-emulation software platform ELLA, based on digital models of control system hardware and software elements supported by virtual reality environment. ELLA consists of many subsystems, e.g.: monitoring and control system WATCH for mobile vehicles control, modular production system, pattern recognition subsystem, quality control system with laser measurement and control units, etc. Figure shows the framework of ELLA simulation & emulation environment.

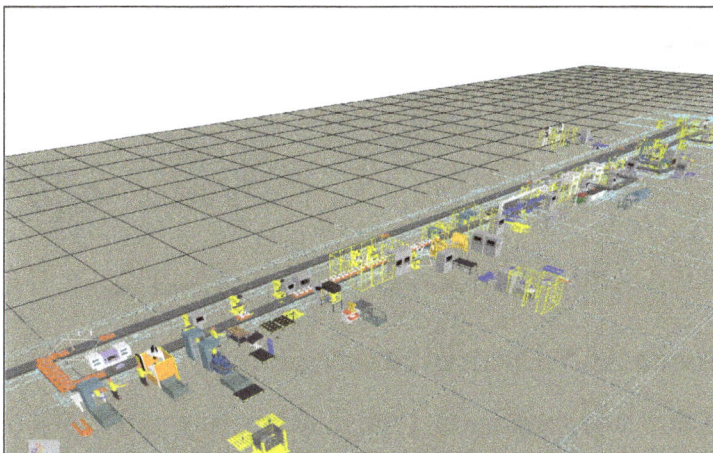

VW Slovakia – FMU of Gearbox Assembly Line.

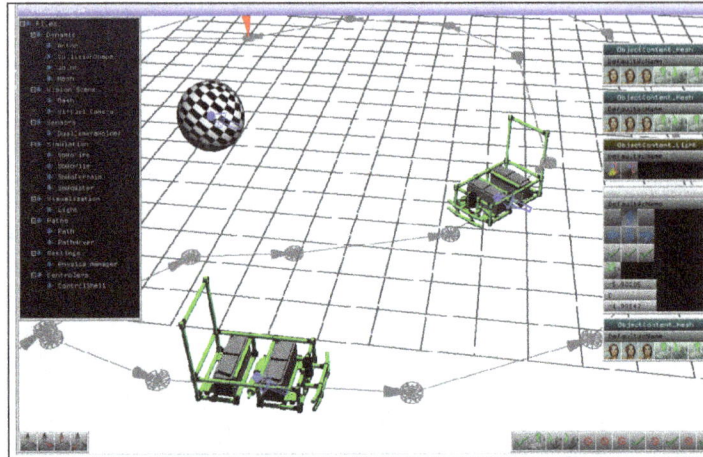

ELLA – Simulation & Emulation Frontend Platform.

Material transportation and handling system belongs among the most decisive parts of effective production systems. Future production systems will require Intelligent Automatic Handling System (IAHS) equipped with industrial robots, autonomous mobile robotic systems, etc. Recent development on this area showed the growing interest in AGVs (Automated Guided Vehicles). There exists a plenty of solutions for automatic transportation and handling of material in production, e.g.: inductive AGVs, through magnetic type controlled AGVs, radio frequency controlled AGVs, mechanically (hanged systems) controlled AGVs, AGVs with artificial intelligent control, etc.

CEIT has developed the platform of low cost AGV solutions for automotive and electronics industries which has been successfully implemented in VW and is currently in testing phase in Continental and Whirlpool factories. The robotic platform control system was fully developed and tested in ELLA environment.

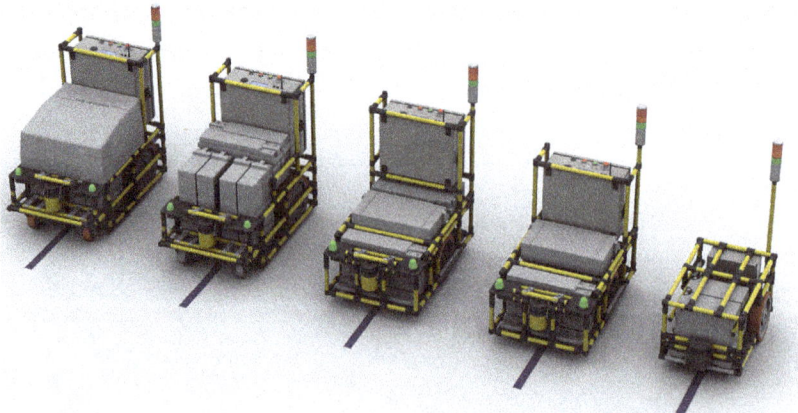

AGV Platform - For Automatic Material Handling.

Simulation of Chosen AGV Control Strategies

The basic principle of simulation resides in a simplified representation of a real (conceptual) system, which we are interested in (simulation target). The analyst does, after verification and validation of a simulation model, a set of simulation experiments. The experimentation with simulation model, aided by a computer, allows examining the variants of system's behaviour in a longer time period and in assumed conditions. New knowledge gained through such simulation experimenting is used for the optimization of a real (conceptual) system.

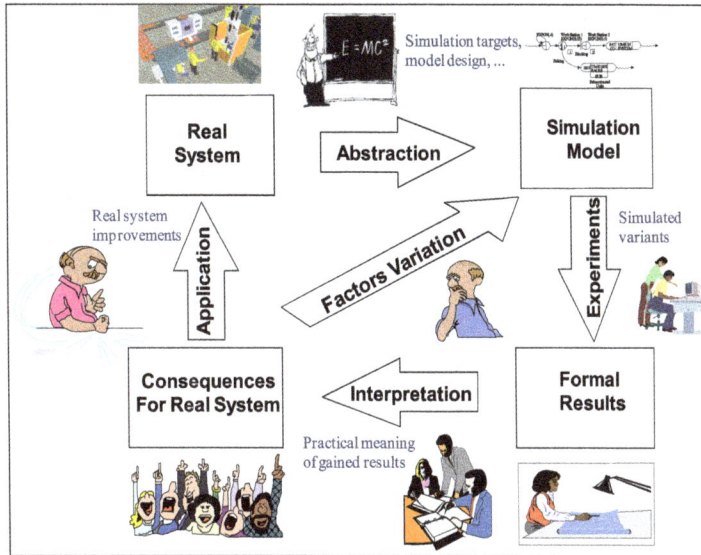

Basic Principle of Computer Simulation.

Simulation is a proven tool for the analysis of dynamic systems. The optimization, especially using Genetic Algorithms and evolutionary approaches, became part of simulation systems.

Chosen AGV control strategies were simulated in ELLA simulation and emulation environment (e.g., FIFO order selection, order selection from the nearest stations, higher priority of the orders in the production, etc.). The simulation project was chosen as a test case for testing of validity of ELLA environment. Figure shows a FMS layout with machines, conveyor, handling robotic workplace, and remote storage area and transportation tracks.

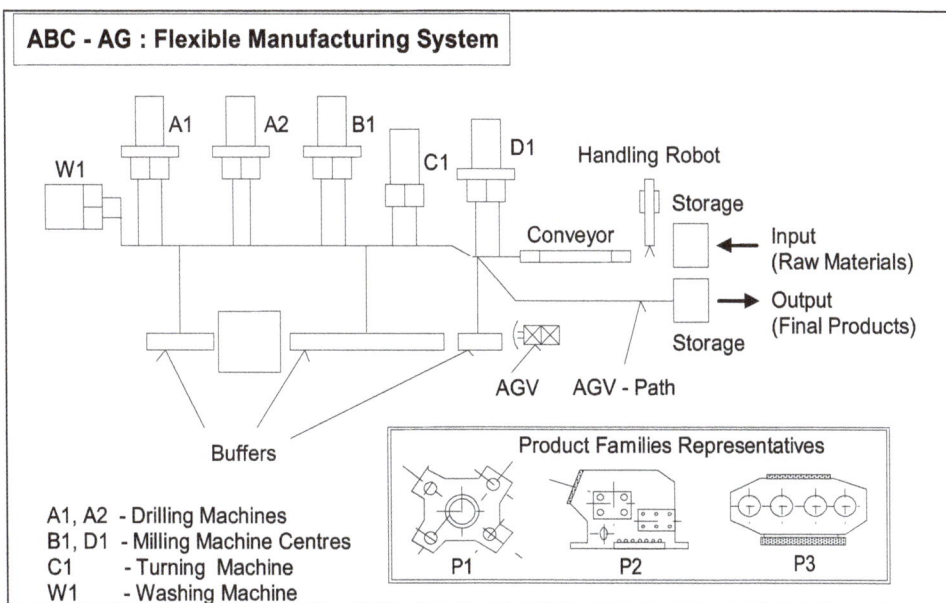

Flexible Manufacturing System Layout.

Following simulation variants were analysed by computer simulation:

Variant 1 - AGV control strategy, FIFO order selection, capacity of I/O buffer on the station – D1.

Variant 2 - AGV control strategy, FIFO order selection, capacity of I /O buffer on the station – C1.

Variant 3 - AGV control strategy, FIFO order selection, capacity of I / O buffer on the station – B1.

Variant 4 - AGV control strategy, FIFO order selection, capacity of I / O buffer on the station – A2.

Variant 5 - AGV control strategy, FIFO, capacity of I / O buffer on the station – A1.

Variant 6 - AGV control strategy, order selection from the nearest station, capacity of I / O buffer on the station – A1.

Variant 7 - AGV control strategy, order selection from the station with the most parts, capacity of I / O buffer on the station – B1.

Variant 8 - AGV control strategy, FIFO, capacity of I / O buffer at the workplaces –A1, B1.

Variant 9 - AGV control strategy, random order selection, capacity of I/O buffer at the workplaces – A1, B1.

Variant 10 - AGV control strategy, random order selection, capacity of I/O buffer at the workplaces – A1, C1.

Variant 11 - AGV control strategy, highest station priority, capacity of I/O buffer at the workplaces – B1, C1.

Variant 12 - AGV control strategy, selection from the nearest station, capacity of I/O buffer on the workplaces - B1, C1.

The results of simulation are shown in figure. Figure shows the progress of Work in Process (WIP) inventory in relation to the production throughput and order average throughput time.

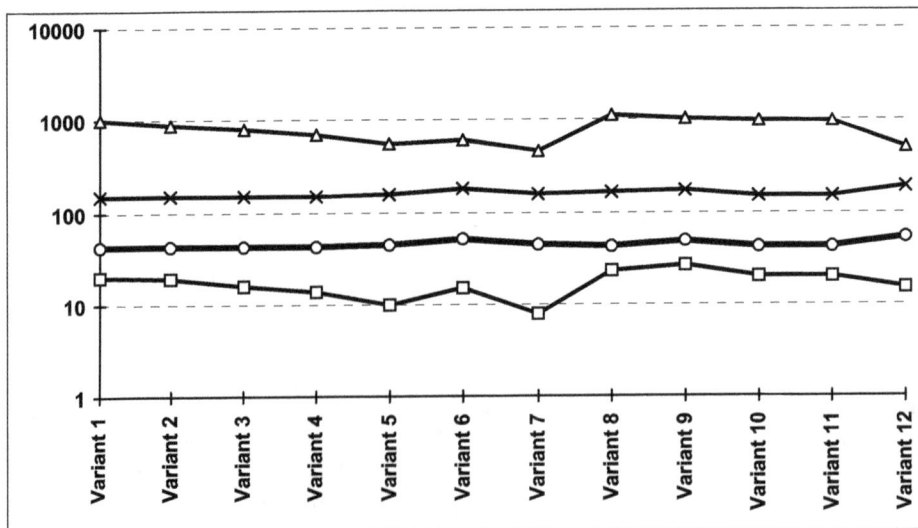

Simulation Results - Chosen AGV Control Strategies.

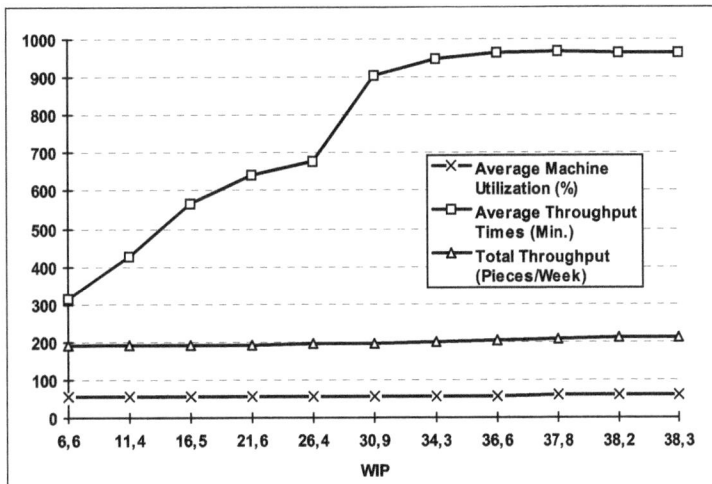

Relationship between WIP and the System Performance (2 AGVs).

The simulation helped to find out the appropriate control strategy and thus it supported the optimization of performance of analysed production system.

Advantages, Restrictions and Benefits of Digital Factory Solutions

Digital Factory implementation results directly in economic as well as production indicators improvement. Any slight saving realised in design and planning phase can bring huge cost reduction in production operation phase. Thanks to this, the payback period of investment in Digital Factory is very short.

Digital Factory Advantages

- Reduction of the entrepreneurial risk by the introduction of a new production,
- Processes verification before start of production,
- Possibility of virtual visit of production halls,
- Validation of designed production concept,
- Optimisation of production equipment allocation,
- Reduction in required area,
- Bottlenecks and collisions analysis,
- Fast changes,
- Better utilization of existing resources,
- Machines and equipment off line programming saving time and resources,
- Reduction or full elimination of prototypes,
- Ergonomics analyses, etc.

Digital Factory enables to test and reveal all possible production problems and shortages before start of production. It enables to eliminate errors in the production line, human or mechanical. Figure shows the main advantages of Digital Factory solutions, e.g. shortening of time to market and significant reduction of lifecycle costs.

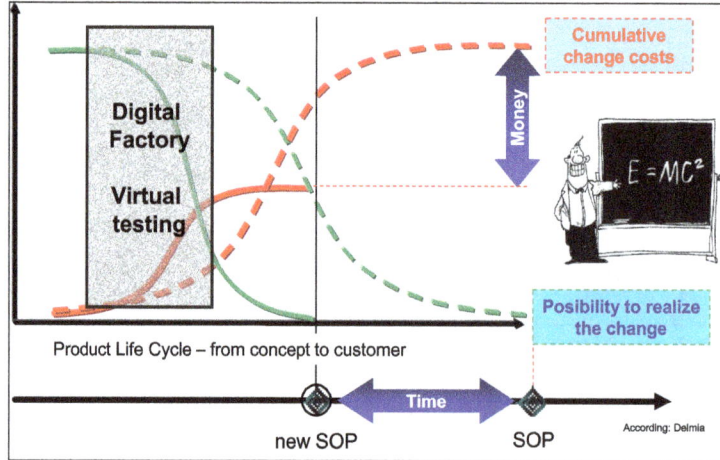

Digital Factory Advantages.

The highest potentials for high quality and low costs of products are in product development and production planning phases. The statistics show that product design and production planning influence about 80% of production costs.

Digital Factory enables product launching time reduction up to 25 - 50%. Estimated cost savings are supposed from 15 to 25%. According to some studies done in industry, using digital manufacturing techniques, twice the amount of design iterations can be processed in 25 percent of the time.

The current production equipment is often inflexible by quick changes. That is why the designers of such equipment are looking for new solutions (automatic reconfiguration of production machines) with fully automated control systems, which will be able to find optimized production process and parameters after production task definition.

According to CIM data (CIM data, 2003), Digital Factory enables to achieve following financial savings:

- Cost savings by assets reduction about 10%.

- Area savings by layout optimisation about 25%.

- Cost savings by better utilisation of resources about 30%.

- Cost savings by material flows optimisation about 35%.

- Reduction in number of machines, tools, workplaces about 40%.

- Total cost reduction about 13%.

- Production volumes growth about 15%.

- Time to market reduction about 30%.

CLOUD MANUFACTURING

Cloud manufacturing (CMfg) is a new manufacturing paradigm developed from existing advanced manufacturing models (e.g., ASP, AM, NM, MGrid) and enterprise information technologies under the support of cloud computing, Internet of Things (IoT), virtualization and service-oriented technologies, and advanced computing technologies. It transforms manufacturing resources and manufacturing capabilities into manufacturing services, which can be managed and operated in an intelligent and unified way to enable the full sharing and circulating of manufacturing resources and manufacturing capabilities. CMfg can provide safe and reliable, high quality, cheap and on-demand manufacturing services for the whole lifecycle of manufacturing. The concept of manufacturing here refers to big manufacturing that includes the whole lifecycle of a product (e.g. design, simulation, production, test, maintenance).

The concept of Cloud manufacturing was initially proposed by the research group led by Prof. Bo Hu Li and Prof. Lin Zhang in China in 2009. Related discussions and research were conducted hereafter, and some similar definitions (e.g. Cloud-Based Design and Manufacturing (CBDM).) to cloud manufacturing were introduced.

Cloud manufacturing is a type of parallel, networked, and distributed system consisting of an integrated and inter-connected virtualized service pool (manufacturing cloud) of manufacturing resources and capabilities as well as capabilities of intelligent management and on-demand use of services to provide solutions for all kinds of users involved in the whole lifecycle of manufacturing.

Types

Cloud Manufacturing can be Divided into Two Categories.

- The first category concerns deploying manufacturing software on the Cloud, i.e. a "manufacturing version" of Computing. CAx software can be supplied as a service on the Manufacturing Cloud (MCloud).

- The second category has a broader scope, cutting across production, management, design and engineering abilities in a manufacturing business. Unlike with computing and data storage, manufacturing involves physical equipment, monitors, materials and so on. In this kind of Cloud Manufacturing system, both material and non-material facilities are implemented on the Manufacturing Cloud to support the whole supply chain. Costly resources are shared on the network. This means that the utilisation rate of rarely used equipment rises and the cost of expensive equipment is reduced. According to the concept of Cloud technology, there will not be direct interaction between Cloud Users and Service Providers. The Cloud User should neither manage nor control the infrastructure and manufacturing applications. As a matter of fact, the former can be considered part of the latter.

In CMfg system, various manufacturing resources and abilities can be intelligently sensed and connected into wider Internet, and automatically managed and controlled using IoT technologies (e.g.,

RFID, wired and wireless sensor network, embedded system). Then the manufacturing resources and abilities are virtualized and encapsulated into different manufacturing cloud services (MCSs), that can be accessed, invoked, and deployed based on knowledge by using virtualization technologies, service-oriented technologies, and cloud computing technologies. The MCSs are classified and aggregated according to specific rules and algorithms, and different kinds of manufacturing clouds are constructed. Different users can search and invoke the qualified MCSs from related manufacturing cloud according to their needs, and assemble them to be a virtual manufacturing environment or solution to complete their manufacturing task involved in the whole life cycle of manufacturing processes under the support of cloud computing, service-oriented technologies, and advanced computing technologies.

Four types of cloud deployment modes (public, private, community and hybrid clouds) are ubiquitous as a single point of access.

- Private cloud refers to a centralized management effort in which manufacturing services are shared within one company or its subsidiaries. Enterprises' mission-critical and core-business applications are often kept in a private cloud.

- Community cloud is a collaborative effort in which manufacturing services are shared between several organizations from a specific community with common concerns.

- Public cloud realizes the key concept of sharing services with the general public in a multi-tenant environment.

- Hybrid cloud is a composition of two or more clouds (private, community or public) that remain distinct entities but are also bound together, offering the benefits of multiple deployment modes.

Resources

From the resource's perspective, each kind of manufacturing capability requires support from the related manufacturing resource. For each type of manufacturing capability, its related manufacturing resource comes in two forms, soft resources and hard resources.

Soft Resources

- Software: Software applications throughout the product lifecycle including design, analysis, simulation, process planning, and are only beginning to be embraced by the electronics manufacturing industry.

- Knowledge: Experience and know-how needed to complete a production task, i.e. engineering knowledge, product models, standards, evaluation procedures and results, customer feedback, and manufacturing in the cloud provides just as many solutions as the number of questions it also raises for manufacturing executives wanting to make the best possible decision.

- Skill: Expertise in performing a specific manufacturing task.

- Personnel: Human resource engaged in the manufacturing process, i.e. designers, operators, managers, technicians, project teams, customer service, and etc.

- Experience: Performance, quality, client evaluation and etc.

- Business Network: Business relationships and business opportunity networks that exist in an enterprise.

Hard Resources

- Manufacturing Equipment: Facilities needed for completing a manufacturing task, e.g. machine tools, cutters, test and monitoring equipment and other fabrication tools.

- Monitoring/Control Resource: Devices used to identify and control other manufacturing resource, for instance, RFID (Radio-Frequency IDentification), WSN (Wireless Sensor Network), virtual managers and remote controllers.

- Computational Resource: Computing devices to support production process, e.g. servers, computers, storage media, control devices, and etc.

- Materials: Inputs and outputs in a production system, e.g. raw material, product-in-progress, finished product, power, water, lubricants, and etc.

- Storage: Automated storage and retrieval systems, logic controllers, location of warehouses, volume capacity and schedule/optimization methods.

- Transportation: Movement of manufacturing inputs/outputs from one location to another. It includes the modes of transport, e.g. air, rail, road, water, cable, pipeline and space, and the related price, and time taken.

CYBER MANUFACTURING

Cyber manufacturing is a transformative system that translates data from interconnected system into predictive and prescriptive operations to achieve resilient performance. It intertwines industrial big data and smart analytics to discover and comprehend invisible issues for decision making. In addition, a cyber-physical interface (CPI) plays a key role in cyber security for connected machines.

Cyber manufacturing evolved from e-manufacturing, which is a systematic methodology that enables manufacturing operations to successfully integrate with the functional objectives of an enterprise through the use of tether-free (i.e. Internet, wireless, web, etc.) communication and predictive technologies. e-manufacturing and cyber manufacturing aim to reduce unexpected downtime and integrate operations with enterprise objectives. However, cyber manufacturing is targeting a system that is much more complex and data- rich, where technologies including smart analytics, distributed systems, control science, and operation management need to be integrated to construct a cyber-physical model. Also, unlike problem-specific solutions in e- manufacturing, cyber manufacturing deploys digital twin technologies to support the life cycle of products. This will enable control systems to compensate or responsible personnel to intervene at the right time on the right assets. Currently, GE is developing Brilliant Manufacturing as a transformational system in the cyber manufacturing environment.

Challenges

Lack of Standards for Seamless Connectivity

Compared to existing Internet-enabled industries, manufacturing assets are less connected, and even those assets that are tend to follow customized protocols. In spite of progresses made in CNC machines, which resulted in protocols such as MTConnect, the majority of the equip- ment use different types of sensors, hardware and software, which leads to different data formats and acquisition requirements. Such situations leave end-users faced with challenges to bring seamless connectivity into their manufacturing plants. Currently, there are several initiatives to address standardization issues and to make technologies more open-access. The Digital Manufacturing and Design Innovation Institute (DMDII) has announced an open- source project called Digital Manufacturing Commons (DMC), that enables technology developers to access and interact with data from various vendors. The National Institute of Standards and Technology (NIST) has also endorsed open-source solutions for interoperability of different data sources in a road map of smart manufacturing systems.

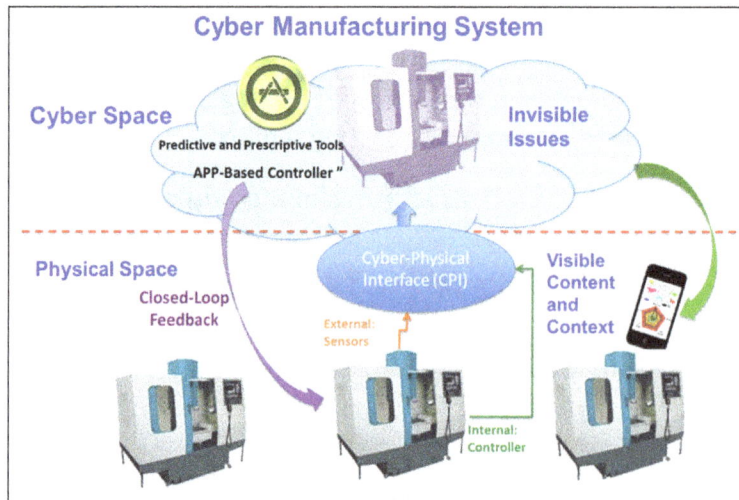

Cyber manufacturing system.

Big Data and Disconnected Analytics

The massive amount of raw data available from factory floor creates opportunities to add intelli- gence to the manufacturing process. McKinsey Report indicated that manufacturing has the largest amount of data stored annually. Meanwhile, the volume, velocity, and variety of the generated data have provided industries with a notice- able challenge: how to extract actionable information from this big data. To address this issue, machine health analytics need to be effectively integrated with factory operation analytics. In addition, technologies such as Hadoop and Spark have been proposed to provide more powerful computational powers through cloud-based and distributed computing.

Industrial Cyber Security

Cyber security is one of the major hurdles in implement- ing cyber manufacturing as it is critical to have resilient capabilities for connected machines and systems in a cloud environment. One approach is to develop a smart cyber- machine interface (CPI) with a time-machine monitoring function so a virtual testing algorithm can be implemented for any control action as input to the physical machine or system. Cyber attacks can be categorized as general and targeted, the latter of which has increased in recent years. Targeted attacks use customized payloads and have higher impacts on the targeted enterprise. Cyber security requires incorporating intelligence-driven approaches instead of solely depending on tools. For example, Defense in Depth is a traditional cyber security strategy that relies on various tools in each layer to ensure protection against attacks. An intelligence-driven use of Defense in Depth requires continuous monitoring of the network and proactively response to potential attacks. Such strategy results in continuous improvement of the defense-intelligence. In addition, developing strict guidelines, technical standards, and educating personnel to avoid social engineering attacks also play a significant role in supporting cyber-security efforts.

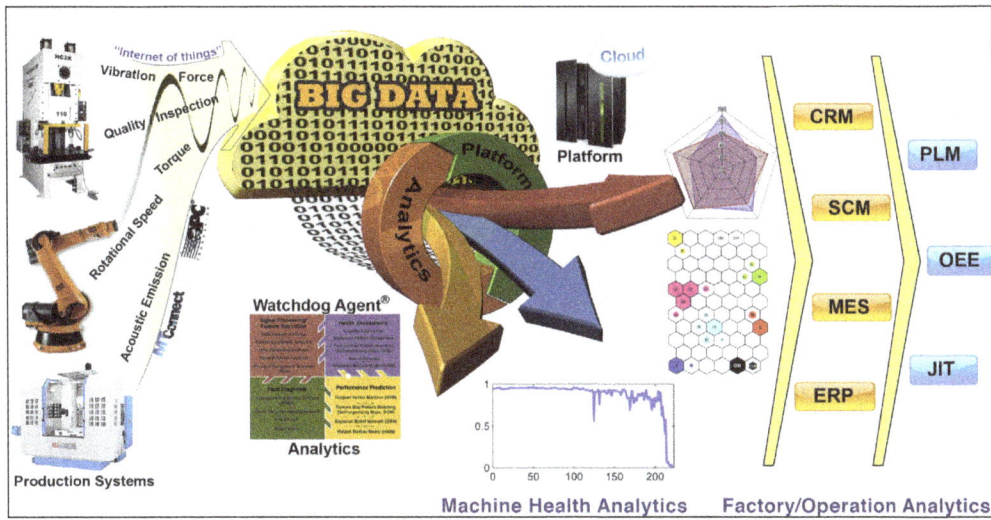

Big data analytics in cyber manufacturing systems.

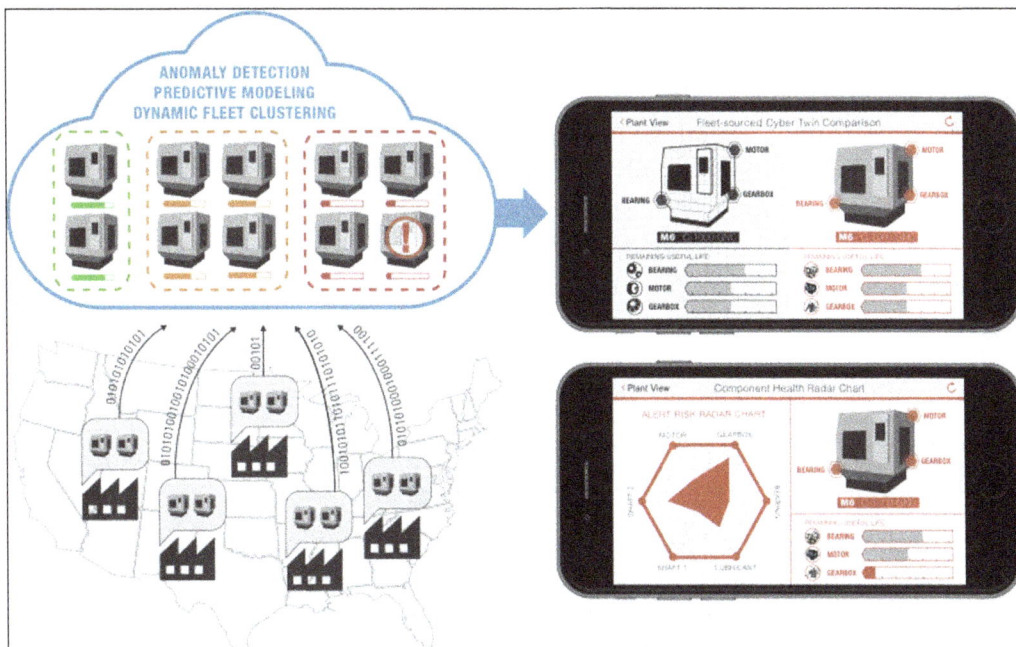

Vision for cyber-physical system-enabled manufacturing.

Enabling Technologies for Cyber Manufacturing

Internet of Things and Predictive Analytics

Internet-of-Things (IoT) defines the universe as the digital threads. The on-going rapid pace of implementing IoT has enabled the manufacturing industry to collect data from an increasing number of manufacturing assets. With semiconductor manufacturing at the frontier, machine tools, band saw machines, conveyors and even products are now able to generate diverse sets of data from controllers and add-on sensors. The key issues in IoT do not stop at connectivity, but also include how to identify critical assets/components to collect the right data, how to synchronize and bridge different sources of data together, and how to conduct analysis. In fact, predictive analytics is what really translates raw data into actionable information and brings out enormous business values of IoT.

Cyber-physical Systems and Platform

The core driving technology of cyber manufacturing is cyber-physical systems (CPS). CPS provides seamless integration between computational models and physical components while offering interoperability and resilience. The "5C" architecture (Connection, Conversion, Cyber, Cognition, and Configuration) indicates that cyber-physical systems will transfer raw data to actionable operations, assist users to comprehend process information, and eventually, to add resilience to the system through evidence-based decision making. Cyber Manufacturing Systems consist of two main functional components: (1) advanced interconnected sensing systems that majorly can be realized through IoT; (2) data management and smart analytics capabilities that transforms raw data into predictive and prescriptive operations. Improved operational processes will then reduce lead time and increase productivity, thus making the enterprise more competitive and robust against changing customer needs.

Transformation to Cyber Manufacturing

As IoT and CPS technologies leap forward, cyber manufacturing realization will be facilitated through standardization between different assets and data hubs, more reliable cyber security, and a general, scalable platform for analytical technologies. Recently, the Center for Intelligent Maintenance Systems has been working on a National Science Foundation-awarded research project on cyber manufacturing. As shown in figure the enterprises with multi-scale, complex, and networked assets will be able to take advantage of cyber manufacturing to maintain productivity by reducing unexpected downtime, reconfigure production assets based on their health status, and thus inject resilience into manufacturing systems.

DIGITAL MATERIALIZATION

Digital materialization (DM) can loosely be defined as two-way direct communication or conversion between matter and information that enables people to exactly describe, monitor, manipulate and create any arbitrary real object. DM is a general paradigm alongside a specified framework that is suitable for computer processing and includes: holistic, coherent, volumetric modeling systems; symbolic languages that are able to handle infinite degrees of freedom and detail in a compact format; and the direct interaction and/or fabrication of any object at any spatial resolution without the need for "lossy" or intermediate formats.

DM systems possess the following attributes:

- Realistic - Correct spatial mapping of matter to information.

- Exact - Exact language and/or methods for input from and output to matter.

- Infinite - Ability to operate at any scale and define infinite detail.

- Symbolic - Accessible to individuals for design, creation and modification.

Such an approach can not only be applied to tangible objects but can include the conversion of things such as light and sound to/from information and matter. Systems to digitally materialize light and sound already largely exist now (e.g. photo editing, audio mixing, etc.) and have been quite effective - but the representation, control and creation of tangible matter is poorly support by computational and digital systems.

Commonplace computer-aided design and manufacturing systems currently represent real objects as "2.5 dimensional" shells. In contrast, DM proposes a deeper understanding and sophisticated manipulation of matter by directly using rigorous mathematics as complete volumetric descriptions of real objects. By utilizing technologies such as Function representation (FRep) it becomes possible to compactly describe and understand the surface and internal structures or properties of an object at an infinite resolution. Thus models can accurately represent matter across all scales making it possible to capture the complexity and quality of natural and real objects and ideally suited for digital fabrication and other kinds of real world interactions. DM surpasses the previous limitations of static disassociated languages and simple human-made objects, to propose systems that are heterogeneous, interacting directly and more naturally with the complex world.

Digital and computer-based languages and processes, unlike the analogue counterparts, can computationally and spatially describe and control matter in an exact, constructive and accessible manner. However, this requires approaches that can handle the complexity of natural objects and materials.

DISTRIBUTED MANUFACTURING

Distributed manufacturing also known as distributed production, cloud producing and local manufacturing is a form of decentralized manufacturing practiced by enterprises using a network of geographically dispersed manufacturing facilities that are coordinated using information technology. It can also refer to local manufacture via the historic cottage industry model, or manufacturing that takes place in the homes of consumers.

Consumer

Within the maker movement and DIY culture, small scale production by consumers often using peer to peer resources is being referred to as distributed manufacturing. Consumers download digital designs from an open design repository website like Youmagine or Thingiverse and produce a product for low costs through a distributed network of 3D printing services such as 3D Hubs or at home with an open-source 3-D printer such as the RepRap.

Enterprise

The primary attribute of distributed manufacturing is the ability to create value at geographically dispersed locations via manufacturing. For example, shipping costs could be minimized when products are built geographically close to their intended markets. Also, products manufactured in a number of small facilities distributed over a wide area can be customized with details adapted to individual or regional tastes. Manufacturing components in different physical locations and then managing the supply chain to bring them together for final assembly of a product is also considered a form of distributed manufacturing. Digital networks combined with additive manufacturing allow companies a decentralized and geographically independent distributed production (Cloud manufacturing).

Social Change

Some call attention to the conjunction of Commons-based peer production with distributed manufacturing techniques. The self-reinforced fantasy of a system of eternal growth can be overcome with the development of economies of scope, and here, the civil society can play an important role contributing to the raising of the whole productive structure to a higher plateau of more sustainable and customised productivity. Further, it is true that many issues, problems and threats rise due to the large democratization of the means of production, and especially regarding the physical ones. For instance, the recyclability of advanced nanomaterials is still questioned; weapons manufacturing could become easier; not to mention the implications on counterfeiting and on "intellectual property". It might be maintained that in contrast to the industrial paradigm whose competitive dynamics were about economies of scale, Commons-based peer production and distributed manufacturing could develop economies of scope. While the advantages of scale rest on cheap global transportation, the economies of scope share infrastructure costs (intangible and tangible productive resources), taking advantage of the capabilities of the fabrication tools. And following Neil Gershenfeld in that "some of the least developed parts of the world need some of the most advanced technologies", commons-based peer production and distributed manufacturing may offer the necessary tools for thinking globally but act locally in response to certain problems and needs. As well as supporting individual personal manufacturing social and economic benefits are expected to result from the development of local production economies. In particular, the humanitarian and development sector are becoming increasingly interested in how distributed manufacturing can overcome the supply chain challenges of last mile distribution.

CLOUD-BASED DESIGN AND MANUFACTURING

Cloud-based design and manufacturing (CBDM) refers to a service-oriented networked product development model in which service consumers are able to configure products or services and reconfigure manufacturing systems through Infrastructure-as-a-Service (IaaS), Platform-as-a-Service

(PaaS), Hardware-as-a-Service (HaaS), and Software-as-a-Service (SaaS). Adapted from the original cloud computing paradigm and introduced into the realm of computer-aided product development, Cloud-Based Design and Manufacturing is gaining significant momentum and attention from both academia and industry.

Cloud-based design and manufacturing includes two aspects: cloud-based design and cloud-based manufacturing. Another related concept is cloud manufacturing that is more general and popular.

Cloud-Based Design (CBD) refers to a networked design model that leverages cloud computing, service-oriented architecture (SOA), Web 2.0 (e.g., social network sites), and semantic web technologies to support cloud-based engineering design services in distributed and collaborative environments.

Cloud-Based Manufacturing (CBM) refers to a networked manufacturing model that exploits on-demand access to a shared collection of diversified and distributed manufacturing resources to form temporary, reconfigurable production lines which enhance efficiency, reduce product life-cycle costs, and allow for optimal resource allocation in response to variable-demand customer generated tasking.

The enabling technologies for Cloud-Based Design and Manufacturing include cloud computing, Web 2.0, Internet of Things (IoT), and service-oriented architecture (SOA).

The term cloud-based design and manufacturing (CBDM) was initially coined by Dazhong Wu, David Rosen, and Dirk Schaefer at Georgia Tech in 2012 for the purpose of articulating a new paradigm for digital manufacturing and design innovation in distributed and collaborative settings. The main objective of CBDM is to further reduce time and cost associated with maintaining information and communication technology (ICT) infrastructures for design and manufacturing, enhancing digital manufacturing and design innovation in distributed and collaborative environments, and adapting to rapidly changing market demands.

In 2014, the same research group also published the worldwide first two books on the subjects of Cloud-Based Design and Manufacturing (CBDM) and Social Product Development (SPD) with Springer, edited by Dirk Schaefer.

Characteristics

CBDM exhibits the following key characteristics:

- Cloud-based distributed file system.

- High performance computing.

- Cloud-based social collaboration.

- Ubiquitous access to distributed big data.

- Rapid manufacturing scalability.

- Agility.

- On-demand self-service.

- Semantic Web.

- Real-time request for quotation.

- Pay-per-use pricing model.

- Multi-tenancy.

CBDM differs from traditional collaborative and distributed design and manufacturing systems such as web-based systems and agent-based systems from a number of perspectives, including (1) computing architecture, (2) data storage, (3) sourcing process, (4) information and communication technology infrastructure, (5) business model, (6) programming model, and (7) communication.

Service Models

- Infrastructure as a service (IaaS).

- Platform as a service (PaaS).

- Hardware as a service (HaaS).

- Software as a service (SaaS).

Similar to cloud computing, CBDM services can be categorized into four major deployment models: the public cloud, private cloud, hybrid cloud, and community cloud.

Digital Modeling and Fabrication

Digital modeling and fabrication is a design and production process that combines 3D modeling or computing-aided design (CAD) with additive and subtractive manufacturing. Additive manufacturing is also known as 3D printing, while subtractive manufacturing may also be referred to as machining, and many other technologies can be exploited to physically produce the designed objects.

Modeling

Digitally fabricated objects are created with a variety of CAD software packages, using both 2D vector drawing, and 3D modeling. Types of 3D models include four models wireframe, solid, surface and mesh. A design is having one or more of these model types.

Machines for Fabrication

Three machine are popular for fabrication:

- CNC router.

- Laser cutter.

- 3D Printer.

CNC Milling Machine

CNC stands for Computer Numerical Control. CNC mills or routers include proprietary software which interprets 2D vector drawings or 3D models and converts this information to a G-code, which represents specific CNC functions in alphanumeric format which the CNC mill can interpret. The G-codes drive a machine tool, a powered mechanical device typically used to fabricate components. CNC machines are classified according to the number of axes that they possess, with 3, 4 and 5 axis machines all being common, and industrial robots being described with having as many as 9 axes. CNC machines are specifically successful in milling materials such as plywood, plastics, foam board, and metal at a fast speed. CNC machine beds are typically large enough to allow 4' × 8' (123 cm x 246 cm) sheets of material, including foam several inches thick, to be cut.

Laser Cutter

The laser cutter is a machine that uses a laser to cut materials such as chip board, matte board, felt, wood, and acrylic up to 3/8 inch (1 cm) thickness. The laser cutter is often bundled with a driver software which interprets vector drawings produced by any number of CAD software platforms.

The laser cutter is able to modulate the speed of the laser head, as well as the intensity and resolution of the laser beam, and as such is able in both to cut and to score material, as well as approximate raster graphics.

Objects cut out of materials can be used in the fabrication of physical models, which will only require the assembly of the flat parts.

3D Printers

3D printers use a variety of methods and technology to assemble physical versions of digital objects. Typically desktop 3D printers can make small plastic 3D objects. They use a roll of thin plastic filament, melting the plastic and then depositing it precisely to cool and harden. They normally build 3D objects from bottom to top in a series of many very thin plastic horizontal layers. This process often happens over the course of a several hours.

Fused Deposition Modeling

Fused deposition modeling, also known as fused filament fabrication, uses a 3-axis robotic system that extrudes material, typically a thermoplastic, one thin layer at a time and progressively builds up a shape. Examples of machines that use this method are the Dimension 768 and the Ultimaker.

Stereolithography

Stereolithography uses a high intensity light projector, usually using DLP technology, with a photosensitive polymer resin. It will project the profile of an object to build a single layer, curing the resin into a solid shape. Then the printer will move the object out of the way by a small amount and project the profile of the next layer. Examples of devices that use this method are the Form-one printer and Os-RC Illios.

Selective Laser Sintering

Selective laser sintering uses a laser to trace out the shape of an object in a bed of finely powdered material that can be fused together by application of heat from the laser. After one layer has been traced by a laser, the bed and partially finished part is moved out of the way, a thin layer of the powdered material is spread, and the process is repeated. Typical materials used are alumide, steel, glass, thermoplastics (especially nylon), and certain ceramics. Example devices include the Formiga P 110 and the Eos EosINT P730.

Powder Printer

Powder printers work in a similar manner to SLS machines, and typically use powders that can be cured, hardened, or otherwise made solid by the application of a liquid binder that is delivered via an inkjet printhead. Common materials are plaster of paris, clay, powdered sugar, wood-filler bonding putty, and flour, which are typically cured with water, alcohol, vinegar, or some combination thereof. The major advantage of powder and SLS machines is their ability to continuously support all parts of their objects throughout the printing process with unprinted powder. This permits the production of geometries not easily otherwise created. However, these printers are often more complex and expensive. Examples of printers using this method are the ZCorp Zprint 400 and 450.

LASER RAPID MANUFACTURING

Selective Laser Melting (SLM) or the closely related Selective Laser Sintering (SLS) differ only in that in SLM complete melting of the powder is achieved as opposed to simply fusing the powder together as happens in the SLS technique. SLM therefore produces fully dense metallic parts with improved mechanical properties.

Laser additive manufacturing (LAM) realizes a multitude of customized medical applications, such as prosthetics or implants. Today, a wide range of biocompatible materials is available for LAM, and yet further processes for metals, plastics or ceramics are under development. Moreover, due to the technology's feasibility to cost-effectively produce lot sizes of one combined with its high degree of freedom of design, products can be tailored to exactly meet a patient's need, such as exact geometrical fit, specific load-bearing behavior or bioresorbability. A tailored product design is created through a digital reverse engineering process that is based on the recorded patient's data. Such customized products include artificial joint or bone replacements, instruments for surgery or prosthetics. They all comprise to enhance the patient's situation and treatment outcome.

Laser additive manufacturing techniques are increasingly being used for biomedical applications; examples of clinical applications are individualized laser-sintered titanium implants and scaffolds for bone regeneration. In research, laser-generated scaffolds are generated from many different materials and seeded with cells to develop cartilage or bone models. For some years, three-dimensional printing has also been applied for printing the cells themselves to generate biological tissue. The ultimate aim of this research is printing complete functional organs. This aim is currently far away being realized, but promising initial studies have demonstrated successful printing of three-dimensional cell models and simple tissue constructs. Cell printing poses specific requirements on the printing technique regarding temperature, humidity, pH value, and other physiological and rheological conditions. Laser-assisted bioprinting based on the laser-induced forward technique meet these requirements. Here, the printing technique, specific cell printing conditions, and applications realized to date are presented.

Virtual Manufacturing

Virtual Manufacturing is nothing but manufacturing in the computer. Presently, Virtual Manufacturing has no unique/standard definition. But few definitions proposed on VM are as follows:

VM is a concept of executing manufacturing processes in computers as well as in the real world, where virtual models allow for prediction of potential problems for product functionality and manufacturability before real manufacturing occurs by Professor Gloria J Wiens. VM is the integrated application of simulation, modeling and analysis technologies and tools to enhance manufacturing design and production decisions and control at all process levels by Professor Edward Lin. VM is a map of practical manufacturing process on computer that is to conduct cooperative work on computer by applying computer simulation and virtual reality technique under support of high-performance computer and super-speed network and to realize product design, art planning, processing manufacturing, performance analysis, quality inspection and essential process of product manufacturing in all-level process management and control in enterprises so as to enhance all decision-making and control capacity in manufacturing process offered comparatively complete definition to current research and developing situation of manufacture. Ka Iwata defines VM as "manufacture of virtual products defined as an aggregation of computer-based information that provide a representation of the properties and behaviours of an actualized product." Some researchers present VM with respect to virtual reality (VR). On one hand, Bowyer A represents VM as a virtual world for manufacturing, on the other hand, one can consider virtual reality as a tool which offers visualization for VM. The most comprehensive definition has been proposed by the Institute for Systems Research, University of Maryland, and discussed by Lin E is "an integrated, synthetic manufacturing environment exercised to enhance all levels of decision and control." "Virtual Manufacturing is the use of a desktop virtual reality system for the computer aided design of components and processes for manufacturing - for creating viewing three dimensional engineering models to be passed to numerically controlled machines for real manufacturing". This definition emphasizes the functions aiding the machining process.

Virtual Manufacturing is the advanced form of Computer Aided Manufacturing based on Virtual reality and Augmented Reality. The concept of artificial reality already appeared in 1970s by Miron Krueger and with the notion of this virtual reality concept was introduced by Jaron Lanier in 1989. In 1990 the concepts of Virtual World and Virtual Environments appeared. Virtual reality is defined as a computer generated interactive and immersive 3D environment simulating reality.The term VM first came into prominence in the early 1990s, in U.S. Department of Defense Virtual Manufacturing Initiative. Both the concept and the term have recently gained wide international acceptance and broadened in scope. For the first half of the 1990s, pioneering work in this field has been done by a handful of major organizations, mainly in the aerospace, earthmoving equipment, and automobile industries, plus a few specialized academic research groups. Recently accelerating worldwide market interest has become evident, fueled by price and performance improvements in the hardware and software technologies required and by increased awareness of the huge potential of virtual manufacturing. Virtual manufacturing can be considered one of the enabling technologies for the rapidly developing information technology infrastructure. In product manufacturing techniques and organization, virtual reality has become the basis of virtual manufacturing aimed at meeting the expectations of the users/buyers of products, also as to their low cost and lead time. Virtual manufacturing includes the rapid improvement of manufacturing processes without drawing on the machines operating time fund.

Types of VM

Virtual manufacturing can be categorized into three groups.

Type of Product and Process Design

Design-oriented Virtual Manufacturing.

During the design phase design oriented VM provides manufacturing information to the design engineer. The parts or the whole machine are simulated and evaluated to test the manufacturability and assembly ability. The purpose is to optimize product design and process design through manufacturing simulation to accomplish manufacturing goals such as Design for Assembly (DFA), quality, flexibility.

- Production-oriented Virtual Manufacturing.

  o Simulation technology is used in production planning or new process model to evaluate and optimize manufacturing processes based on Integrated Product Process Development and test the rationality and producibility of process flow.

- Control-oriented Virtual Manufacturing.

  o Analysis technology is applied to control model to simulate production management activities of production line or workshop so as to realize optimal control based on models and actual process.

Type of system integration.

According to the definitions proposed by Onosato and Iwata, every manufacturing system can be decomposed into two different sub-systems:

- Real Physical System (RPS).

  o An RPS is composed of substantial entities such as materials, parts and machines that exist in the real world.

- Real Informational System (RIS).

o   An RIS involves the activities of information processing and decision making.

- Virtual Physical System (VPS).

    o   A computer system that simulates the responses of a real physical system is a virtual physical system, which can be represented by a factory model, product model, and a production process model. The production process models are used to determine the interactions between the factory model and each of the product models.

- Virtual Information System (VIS).

    o   A computer system that simulates a RIS and generates control commands for the RPS is called a 'virtual- informational system (VIS).

Type of Functional Usage

VM is used in the interactive simulation of various manufacturing processes such as virtual prototyping, virtual machining, virtual inspection, virtual assembly and virtual operational system.

- Virtual Prototyping (VP) mainly deals with the processes, tooling, and equipment such as injection molding processes. VM is allied to the Virtual Prototyping, the Virtual CAD and Virtual CAM made most of the time by simulation. Roger W Pryor discussed in his paper on the potential real benefits that can be realized through cost saving, minimization of number of prototype models.

- Virtual machining mainly deals with cutting processes such as turning, milling, drilling and grinding, etc. The VM technology is used to study the factors affecting the quality, machining time and costs based on modeling and simulation of the material removal process as well as the relative motion between the tool and the workpiece.

- Virtual inspection makes use of the VM technology to model and simulate the inspection process, and the physical and mechanical properties of the inspection equipment.

- In Virtual Assembly, VM is mainly used to investigate the assembly processes, the mechanical and physical characteristics of the equipment and tooling, the interrelationship among different parts and the factors affecting the quality based on modeling and simulation. A virtual assembly environment would enable a user to evaluate parts that are designed to fit together with other parts. Issues such as handling ease of assembly and order of assembly can be studied with virtual assembly.

- Virtual operational control makes use of VM technology to investigate the material flow and information flow as well as the factors affecting the operation of a manufacturing system.

Vision of VM

The vision of Virtual Manufacturing is to provide a capability to make it in the computer. VM provides a modeling and simulation environment such that the design, analysis and fabrication/assembly of any product including the associated manufacturing processes which can be simulated in the computer. Figure represents the vision on VM.

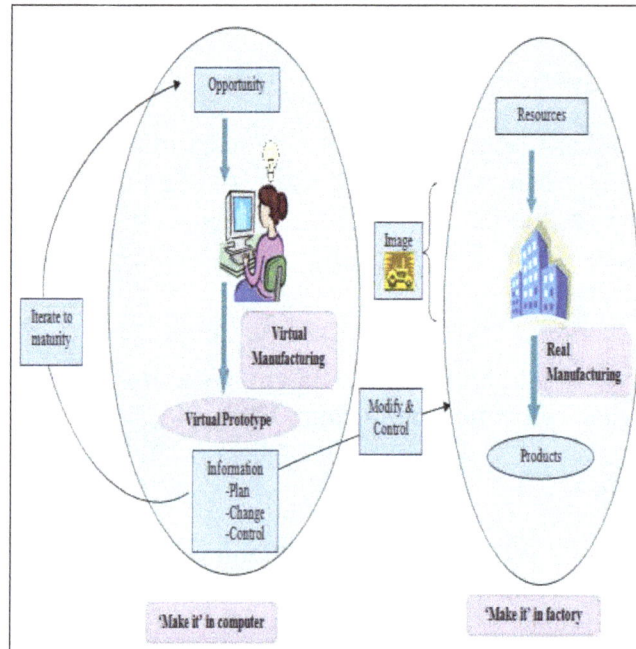

Vision of Virtual Manufacturing.

The needs for VM are technology and development for models and simulations, virtual infrastructures and tools for working in virtual spaces such as CAM/CAM. The scope of VM varies from workshop practice to company areas. To establish a theoretical approach of VM, knowledge of CAD and Virtual Prototyping (VP) is essential. VM characterizes virtualization of the real to the realization of the virtual.

Benefits of VM

Virtual Manufacturing builds confidence to manufacturers of knowing the delivery of quality products to market on time and within the initial investment. The benefits of VM from product point of view, it will improve quality of the product, reduce number of physical prototype models, allows simulations of multiple manufacturing products, optimize the design of product and processes. And from production point of view, it will improve the confidence in the process, reduce material waste, reduce tooling cost, lowers manufacturing cost and optimizes manufacturing processes .

Implementing VM contributes several benefits such as ensuring higher quality of the tools, lesser cycle time for production of parts without false start, optimize the design for manufacturing and assembly system for better producibility, flexibility in mix production of multiple products, quick response to customers about the impact of investment and delivery schedule with improved accuracy, good relations with the customers.

Virtual Manufacturing Systems

VM integrates manufacturing activities dealing with models and simulations, instead of objects and their operations in the real world. VM systems produce digital information to facilitate physical manufacturing processes. The concept, significance, and related key technology of VM were addressed by Lawrence Associate Inc. while the contribution and achievements of VM were overviewed by Shukla. Kimura explained a typical VM system consists of a manufacturing resource model, a manufacturing environment model, a product model, and a virtual prototyping model. Virtual manufacturing systems are synthetic environment designed to exhibit manufacturing systems operation on a reality virtuality continuum. The reality–virtuality state exhibited by the manufacturing system falls into the following categories.

- Reality: Real manufacturing operation.

- Augmented reality: Manufacturing system control is augmented by the use of electronic hardware and computer software in order to facilitate managers with more micro- and macro-level parameters for accurate decision making leading to higher profitability.

- Augmented virtuality: Consist of higher level of virtuality than augmented reality. In augmented virtuality, a higher proportion of elements are synthetic in nature.

- Virtuality: Encompasses immersion in a completely synthetic environment. The integration of virtual reality technology into the conceptual design and process planning stages reduces design time and cost.

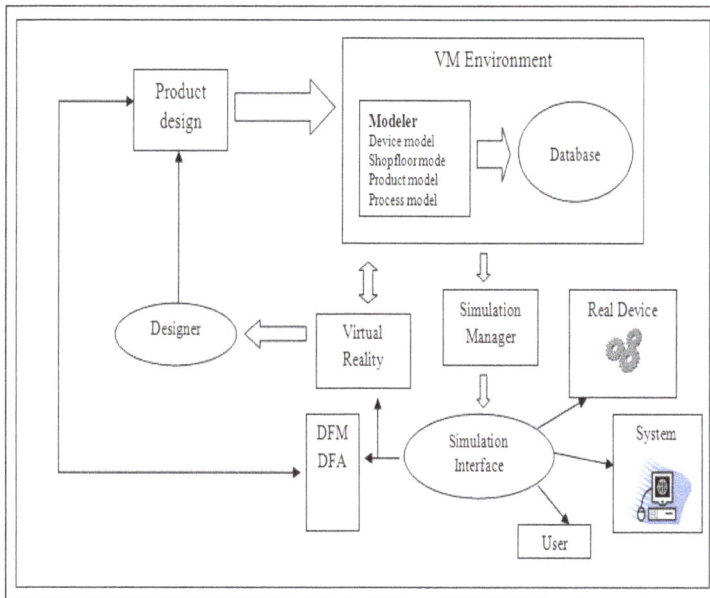

Virtual Manufacturing Systems.

Virtual manufacturing system as shown in figure is a representation of the complex real manufacturing system, consisting of various interacting, interrelated, and interdependent sub-systems and processes represented by abstract models.

Virtual manufacturing systems focused on specific applications like geometric modeling, product modeling, production system modeling, etc. These systems are very complicated, not modularized, not easy to construct, calibrate, or validate, not adaptable, and not easy to reuse. One of the possible reasons for this is that most of the system models are build hysterically, without first identifying the logical system structure. Therefore, for the purpose of construction of a comprehensive VM system, an unmistakable system structure, based on the various aspects of the real production process, is the necessary prerequisite.

The system structure helps in identifying the interactions, interrelations and functioning of various elements of the whole VM system. The interpretation of the dynamic relationships between various system elements facilitates the validation, calibration and verification of the individual system components as well as the whole VM system.

Virtual manufacturing must cope with many models of various types of product and process and require a large amount of computation simulating the behavior of equipment on a shop floor. In order to deal with this complexity in manufacturing, it is necessary to approach an open system architecture of modeling and simulation for virtual manufacturing systems.

K Iwata developed a modeling and simulation framework for virtual manufacturing systems, a modeling method integrating various simulation resources, and a control mechanism for distributed simulation. Based on K. Iwata's open system architecture, H. Hanwu explained a new modeling and simulation architecture for virtual manufacturing system. In H Hanwu and co-authors opinion, modeling means to set up a virtual manufacturing environment and simulation means to achieve virtual manufacturing activities. Figure represents the outline of VM environment.

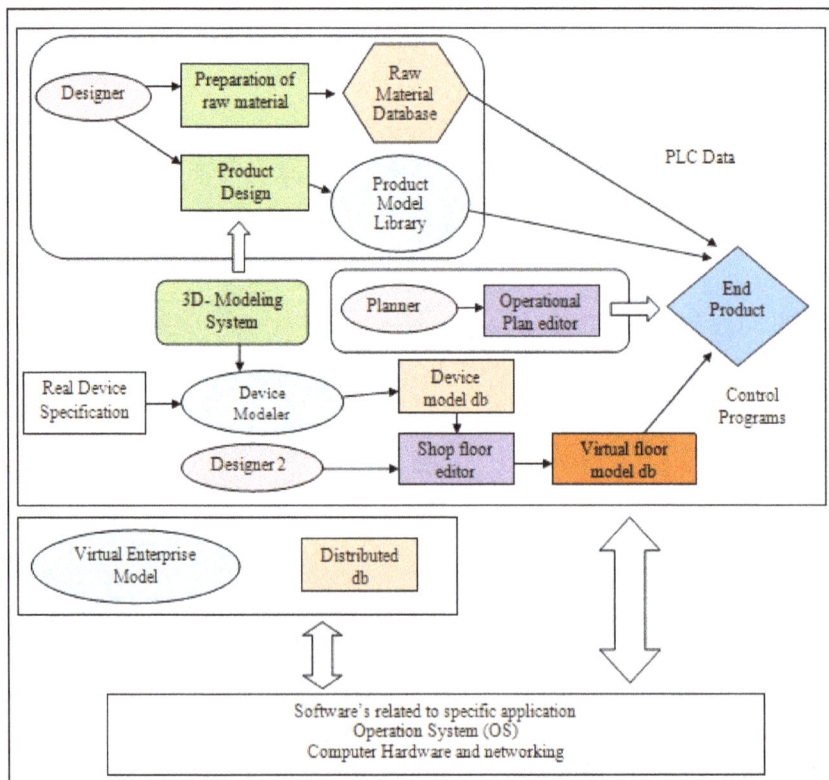

Virtual Environment.

Scott Angster describes a system named as Virtual Environments for Design And Manufacturing (VEDAM), that has been designed and partially implemented to support virtual design, virtual manufacturing and virtual assembly. VEDAM also aimed at extending the capabilities of existing parametric CAD/CAM systems. Jayaram discuss the initial requirements for a virtual manufacturing environment (VME) as a part of an applications development environment. Rosen, D.W expressed in his paper that parts that are being designed using Virtual Design Studio will be disassembled within the virtual environment to identify and correct re-manufacturing and servicing problems.

The following task should be accomplished to construct the virtual manufacturing environment:

Product model library: It corresponds to product modelling and provides the initial models of workpieces for simulating production process. It prepares product data used for fabricating products such as Computer Numerical Controlled (CNC) programs for machining and measuring machines.

Device model preparation: In order to calculate the behavior of devices, device model database are developed as per the real device specification.

Virtual shopfloor model construction: Virtual factory is defined by the construction of virtual shopfloor which uses device library models.

Virtual enterprise organization: A virtual enterprise is a partnership of companies that forms in response to a certain market opportunity. Distributed manufacturing is performed by virtual enterprises. The associates, who may geographically distributed and of various sizes, technical sophistication, contribute their core competence to enterprise, enhancing its ability to deliver high quality. Wei Cui proposed the service-oriented architecture of Virtual Enterprise for manufacturing industry. Yiru Dai, Juanwei Yan describes the modeling of agile manufacturing enterprises plays an important role in agile manufacturing and the important position in life circle of Virtual Enterprise.

VirtuE is a VE model used in information markets to provide cooperation,amoung distributed architecture, transactions between members and methods of measuring performance and controlling behavior.

Methods and Simulation Tools used in Virtual Manufacturing Systems

VM has two main core activities. The first one is the "Modeling Activity" which determines what to model and degree of thought that is needed. The second on is the "Simulation Activity" which represents model in a computer based environment and compare to the response of the real system with degree of accuracy and precision.

The following methods are necessary to achieve VM system:

Manufacturing characterization confines measure and analyze the variables that influence material transformation during manufacturing. Modeling and representation technologies provide different kinds of models for representation, standardization the processes in such a way that the information can be shared between all software applications (Knowledge based systems, Object oriented, feature based models). Visualization, environment construction technologies includes Virtual reality techniques, augmented reality technology, graphical user interfaces for representation of information to the user in a meaningful manner and easily comprehensible. Verification, validation and measurement the tools and methodologies needed to support the verification and validation of a virtual manufacturing system. Multi discipline optimization: VM and simulation are usually no self-standing research disciplines, they often are used in combination with "traditional" manufacturing research. Nowadays numerous tools are available for simulating manufacturing levels.

Applications of VM

The virtual manufacturing has been successfully applied to many fields such as, automobile manufacturing, aeronautics and astronautics, railway locomotives, communication, education and so on, which has an overpowering influence on industrial circles.

Automotive Domain

The shape design using the virtual technology can be modified and evaluated at any time. The modeling data after scheme confirming can be directly used for the stamping tool design, simulation and processing, even for the marketing and propaganda. Application of VM is used in automobile factory shop floor and also in car driving simulation. Song Cheng describes a case research of D auto-company's virtual paint shop established with the technology of three dimensional simulations. Song Cheng depicted various equipments, structural objects, the way of establishing model, concerning contents and the simulation process of the whole shop and provide relatively real virtual shop and equipments' datum for the engineers.

Aerospace Domain

Virtual Manufacturing in aerospace industry is used in FEA to design and optimize parts, e.g. reduce the weight of frames by integral construction, in 3D-kinematics simulation to program automatic riveting machines, and few works dealing with augmented reality and virtual reality to support complex assembly and service tasks in aircraft design. The aero engine model created in virtual environment describes where tools are developed and used to help manufacturing and design engineers to take action and decisions on problems normally solved only by experience. Henrik R explained application of VM in aircraft domain by considering Turbine Exhaust Casing (TEC). TEC is manufactured by fabrication and about 200 welds are needed to manufacture the product. Issues have been identified with the robustness of the geometrical tolerances created during production. Several welding sequence concepts were investigated to find a more robust manufacturing sequence. From the welding simulations it was shown that the residual stresses could be lowered using a different welding sequence. Moreover, to further avoid the issue with geometrical tolerances a pre-deformation was given to the product before welding, the amount of needed pre-deformation was calculated by the virtual welding simulation tool.

Healthcare Domain

Healthcare is one of the biggest adopters of virtual reality which encompasses surgery simulation, phobia treatment, robotic surgery and skills training. One of the advantages of this technology is that it allows healthcare professionals to learn new skills as well as refreshing existing ones in a safe environment. Plus it allows this without causing any danger to the patients. Virtual manufacturing applications in the healthcare industry are associated with many leading areas of medical technology innovation including robot-assisted surgery, augmented reality (AR) surgery, computer-assisted surgery (CAS), image-guided surgery (IGS), surgical navigation, multi-modality image fusion, medical imaging 3D reconstruction, pre-operative surgical planning, virtual colonoscopy, virtual surgical simulation, virtual reality exposure therapy (VRET), and VR physical rehabilitation and motor skills training. Stent design influences the post-procedural hemodynamic and solid mechanical environment of the stented artery by introducing non-physiologic flow patterns and elevated vessel strain. This alteration in the mechanical environment is known to be an important factor in the long-term performance of stented vessels. Because of their critical function, stent design is validated by methods such as FEA.

Home Appliance Domain

The virtual kitchen equipment system developed by a Japanese company Matsushita allows customers to experience functions of a variety of equipment in virtual kitchen environment before the purchase of actual equipment. These choosing results can be stored and send to the production department through computer network and be manufactured.

Other Applications of VM Explicated by Li Liu

Product shape style designs of conventional automobiles adopt the plastic to manufacture the shape model. The shape design using the virtual technology can be modified and evaluated at any time. In the shape design of other products such as building and decoration, cosmetic packing, communication, etc. has great advantages. In piping system design, through the implementation of virtual technology, the designer can enter into virtual assembly by conducting piping layout and check the potential interference and other problems. Product movement and dynamics simulation displays the product behavior and dynamically perform the product performance. The product design must solve the movement coordination and cooperation of each link on the production line. The usage of simulation technology can intuitively conduct the configuration and design, and guarantee the working coordination. In product assembly simulation the coordination and assembly property of mechanical product is the place where most errors of the designers emerge. In the past, the error at final stage leads to the scrapping of parts and delay manufacture product which causes more economic losses and damage.

The implementation of virtual assembly technology can conduct the verification in the stage of design, and ensure the correctness of design to avoid the loss. The adoption of virtual reality technology in virtual prototype suitably helps in 3D modeling of products, and then set the model into VE to control, simulate and analyze. Simulation and optimization of the productive process of enterprise are used in the productive technology by formulating the products, man power of the factory, reasonable allocation of manufacturing resources, material storage and transportation system. LIU Qing-ling addressed the VM system provides the working environment of collaboration for the virtual enterprise partners, that affords collaboration support for each link of the whole course of orders of users, originality in product, design, production of parts, set assembling, sales and after sale services. Virtual Simulation is an important technologic method accounting for complex design and testing of designing proposal. Yongkang Ma explains in his research that the elements such as welding robots and fixtures of workstation for body-in-white welding are analysed and optimized using digital modelling method of work-station .

Virtual Teaching Platform of Digital Design and Manufacturing

To promote students' learning interest and improve teaching effects researchers adopts a virtual teaching platform of digital design and manufacturing in innovation teaching methodology. Yu Zhang explains virtual reality technology in program -based learning helps students to establish their spatial concepts and enhance their understanding on engineering drawings. Huang Xin represents motion simulation of entire product mechanisms could be achieved by means of the function of intelligent simulation. Liu Jianping suggests that with the help of the CAD software, students can easily understand how to read technical drawing and replicate same in software, and the cost of design can also be saved.

Virtual Training

Hazim El-Mounayri concluded that the architecture of a virtual training environment (VTE) was used to develop the corresponding system for the case of CNC milling. A recent application of VE based training includes training for operation of engineering facilities, CNC manufacturing. The Learning Environments Agent (LEA) engine includes a hierarchical process knowledge base engine, an unstructured knowledge base engine for lecture delivery; a rule based expert system for natural-language understanding, and an interface for driving human-like virtual characters. Integrated Virtual Reality Environment for Synthesis and Simulation engine was used to drive the virtual environment, display the engineering facility and manage a multimodal input from a variety of sources. A general geometric modeling approach is based on modeling precisely the geometries involved in the machining operation, including work-piece geometry and tool geometry.

Drawbacks in VM

Firstly, setup of VM system requires huge capital investment for material, simulation software and human. Secondly with respect to availability of simulation model, each time at each level a new model has to be built even though the previous model has been already done. Thirdly compatibility of VM software and hardware is essential for better effectiveness as software depends on latest IT technology. By considering the above mentioned drawbacks, several hot topics are proposed in VM research area:

In human-computer interfaces users expect to interact with the computer in a human like manner. Development of good interfaces not only graphical but also mixtures of text, voice, visual interface are required.

Any kind of planning activity can be supported and improved by simulation. The goal to be reached is to implement integration of simulation systems in planning and design tools so that the benefits can be achieved by minimum effort. Computer Aided Design (CAD) data has to be influenced by automatic generation of simulation models. The goal is to automatically create ready-to-run simulation models out of CAD data with extra information. To make use of adaptation of CAD model to specific need instead of creating a totally new CAD model. A major drawback is the combination of real and simulated hardware in machine tool and manufacturing system development known as Hybrid System Simulation. Real devices, like the machine controller, would be linked to a simulated model of the machine to test the machine's behaviour during manufacturing. The aim of Virtual Protyping is to approach for reliable and very accurate simulation models, which are able to show nearly realistic behaviour under static and dynamic stress conditions. The task of VM is to combine results of different kinds of simulation to predict a nearly realistic behaviour of machine tool, tool and work piece during machining process.

References

- Design of Flexible Production Systems – Methodologies and Tools. By T. Tolio. Berlin: Springer, 2009. ISBN 978-3-540-85413-5

- Whatismanufacturingtechnology, aboutamt: amtonline.org, Retrieved 24 May, 2019

- "Overhead Conveyors and Material Handling". Mhel.co.uk. Archived from the original on 2016-09-02. Retrieved 2016-08-23

- Manufacturing-technology: cerasis.com, Retrieved 25 June, 2019

- Leung, Sonny. "What Is Profile Projector". Vision Measuring Machine and Profile Projector. Retrieved 8 December 2016

- Useful-lean-manufacturing-tools, lean, learn: leankit.com, Retrieved 26 July, 2019

- Ernie Conover (2000), Turn a Bowl with Ernie Conover: Getting Great Results the First Time Around, Taunton, p. 16, ISBN 978-1-56158-293-8

- Material-handling-equipment, materials-handling: thomasnet.com, Retrieved 27 August, 2019

- Griffiths, Tony. "Makers of "Bench Precision" Lathes". LATHES.CO.UK. Archived from the original on 27 December 2017. Retrieved 5 February 2018

INDEX

www.ingramcontent.com/pod-product-compliance
Lightning Source LLC
Chambersburg PA
CBHW082029190326
41458CB00010B/3314